Haldane, Mayr, and
Beanbag Genetics

Haldane, Mayr, and Beanbag Genetics

Krishna Dronamraju

OXFORD
UNIVERSITY PRESS

OXFORD
UNIVERSITY PRESS

Oxford University Press, Inc., publishes works that further
Oxford University's objective of excellence
in research, scholarship, and education.

Oxford New York

Auckland   Cape Town   Dar es Salaam   Hong Kong   Karachi
Kuala Lumpur   Madrid   Melbourne   Mexico City   Nairobi
New Delhi   Shanghai   Taipei   Toronto

With offices in
Argentina   Austria   Brazil   Chile   Czech Republic   France   Greece
Guatemala   Hungary   Italy   Japan   Poland   Portugal   Singapore
South Korea   Switzerland   Thailand   Turkey   Ukraine   Vietnam

Copyright © 2011 by Oxford University Press

Published by Oxford University Press, Inc.
198 Madison Avenue, New York, New York 10016
www.oup.com

Oxford is a registered trademark of Oxford University Press

Library of Congress Cataloging-in-Publication Data
Dronamraju, Krishna R.
  Haldane, Mayr, and beanbag genetics / Krishna Dronamraju.
    p.; cm.
  Includes bibliographical references.
  ISBN 978-0-19-538734-6 (hardcover : alk. paper) 1. Population genetics–History.
2. Evolutionary genetics–History. 3. Haldane, J. B. S. (John Burdon Sanderson),
1892-1964. 4. Mayr, Ernst, 1904-2005. I. Title.
  [DNLM: 1. Haldane, J. B. S. (John Burdon Sanderson), 1892-1964.
2. Mayr, Ernst, 1904-2005. 3. Evolution. 4. Genetics, Population–history.
5. History, 20th Century. QU 11.1 D786h 2011]
QH455.D76 2011
576.5'8–dc22   2010011747

9 8 7 6 5 4 3 2 1
Printed in the United States of America
on acid-free paper

Dedicated to Jim Crow and the late Motoo Kimura

Contents

Introduction

A fact in science, is not a mere fact, but an instance.

—Bertrand Russell

This book is about the views of two famous biologists, both now deceased, and how they differed on a central issue in evolutionary biology. As their correspondence indicates, theirs was a friendly disagreement—there was no animosity or bitterness in their arguments. Indeed, as Ernst Mayr once wrote in a letter to J.B.S. Haldane, it is all a matter of emphasis and interpretation. Their differences reflected their backgrounds and experience.

The term "beanbag genetics" is derived from the fact that the early Mendelians used to keep different colored beans in bags for the purpose of counting and analyzing Mendelian ratios. This method implied that genes behaved as isolated independent entities with no interaction with each other. Mayr (1959, 1963) used the term "beanbag genetics" to describe the methodology and the underlying concepts of early studies in theoretical population genetics by R.A. Fisher (1930), Haldane (1932a), and Sewall Wright (1931). This was especially true of Fisher and Haldane, who used simple models of genes acting in isolation for the sake of mathematical convenience. In his opening address to the 1959 Cold Spring Harbor Symposium on Qualitative Biology "Genetics and Twentieth-Century Darwinism," Mayr challenged the great pioneers of population genetics: "But what, precisely, has been the contribution of this mathematical school to the evolutionary theory, if I may be permitted to ask such a provocative question?"(p.2). It must be emphasized, however, that Mayr erred in lumping Wright with Fisher and Haldane because Wright had always considered the role of interactions of genes (genic interaction) and epistasis in the evolutionary process. In his contribution to *Haldane and Modern Biology*, Wright (1968a, p.5) wrote:

> My theory was directed toward ascertaining whether some way, after all, might exist in which selection could take advantage of the enormous number of interaction systems provided by a limited number of unfixed loci. It was maintained that this is possible in a large population subdivided into many small, local populations, sufficiently isolated to permit considerable random differentiation of gene frequencies, but not so isolated as to prevent gradual diffusion of the more successful interaction systems from their centers of origin.(p.5)

In response to Mayr's challenge, Haldane (1964) wrote a lengthy, witty, irreverent, and pugnacious essay, combining both defense and offense skillfully.

The correspondence between Haldane and Mayr, during the years 1947–1964, is revealing and helpful in understanding their positions on a number of evolutionary topics, including the first issue of the journal *Evolution*, Darwin's contributions, speciation, beanbag genetics, natural selection, genetic loads and the cost of natural selection, butterfly and wasp behavior, and also Haldane's election to the National Academy of Sciences, Mayr's visit to India, Calcutta Zoo, and Haldane's cancer and death in December 1964. One letter was addressed to Konrad Lorenz by Haldane on the subject of sympatric speciation.

These letters between Haldane and Mayr give us glimpses of the nature of their friendship and intellectual camaraderie. In one of his letters (May 8, 1963) to Mayr, Haldane wrote: "I am an unrepentant 'beanbag geneticist.' Nonmathematicians often fail to realize the complexity of behavior, and sometimes the self-regulatory capacity, of material systems composed of simple components." In another letter (April 6, 1963) to Mayr, Haldane wrote: "As a 'beanbag geneticist' I think your view of a species may be a little too physiological and not historical enough. No doubt the various genes (regarding frequency and location as well as molecular pattern) have to fit together." Curious readers will enjoy reading these letters and reach their own conclusions on various subjects.

When Mayr became the first editor of *Evolution* in 1947, he invited Haldane to join the editorial board and sought his advice on editorial policy. They both agreed that the papers should reflect international contributions. Mayr wrote that he was "most anxious" to include some mathematical papers and invited Haldane to submit a paper. The result was Haldane's 1949 paper "Suggestions as to Quantitative Measurement of Rates of Evolution" (*Evolution, 3*: 51–56).

As a young student of Haldane's, I knew both Haldane and Mayr personally. I knew Haldane intimately: Not only did I have daily interactions with him during his last years in India, but we also traveled together to scientific conferences in several countries. And I came to know Mayr well over several decades. His visit to India, when we shared a guest house in Orissa, was especially memorable to me. Haldane had carefully arranged Mayr's visit to the famous temples of Puri and Konark and was delighted to find that Mayr managed to include bird-watching in the mornings.

Several years later, Mayr readily accepted my invitation to contribute a chapter to my 1995 book *Haldane's Daedalus Revisited*, which examined Haldane's futuristic predictions about genetic engineering and eugenics.

Krishna Dronamraju
Foundation for
Genetic Research,
Houston, Texas

Acknowledgments

I am pleased to express my gratitude to the late Ernst Mayr, who graciously gave me permission to reproduce his letters shortly before his death in 2005. I am also grateful to Prof. N.A. Mitchison and Lois Mitchison, J.B.S. Haldane's nephew and niece, respectively, for permitting me to reproduce Haldane's writings in my books and papers published over several years.

Prof. Walter Bock, a close pupil of Mayr's, has kindly provided much information about Mayr's life and work and was especially helpful in preparing the time line of Mayr.

I am most grateful to Gillian Furlong, librarian at University College, London, for help with the J.B.S. Haldane archives, and to Clark A. Elliott, Associate Curator of Harvard University archives, for help with the Ernst Mayr archives. Furthermore, the Rockefeller University archives in Sleepy Hollow, New York, the Chemical Heritage Foundation in Philadelphia, and the Wellcome Trust library in London have been helpful in providing valuable archival information.

It is a pleasure to acknowledge Michele Wambaugh's warm support, encouragement, and assistance while I was writing the book.

J.B.S. Haldane with his associates at the Indian Statistical Institute, Calcutta, in 1961. The author is sitting in the front with J.B.S. Haldane and his former pupil Dr. Pamela Robinson, Paleontologist from University College, London. Courtesy Indian Statistical Institute.

Ernst Mayr in the 1960s. Courtesy Archives of the Ernst Mayr Library of the Museum of Comparative Zoology, Harvard University.

J.B.S. Haldane lecturing in London.

Haldane, Mayr, and Beanbag Genetics

I

What Is Beanbag Genetics?

In 1959, Ernst Mayr questioned the three great population geneticists, R.A. Fisher, Sewall Wright, and J.B.S. Haldane, "But what, precisely has been the contribution of this mathematical school to the evolutionary theory, if I may be permitted to ask such a provocative question?" His question arose in part from the fact that the mathematical theory at the time did not address the subject of speciation, a major interest of Mayr. He focused instead on what he called "beanbag genetics," in which "Evolutionary genetics was essentially presented as an input or output of genes, as the adding of certain beans to a beanbag and the withdrawing of others" (Mayr 1959).

From his many publications on the subject of evolution, it is clear that Mayr had been much troubled by the contribution of geneticists to the understanding of the process of evolution. For instance, in the first chapter of his well known book *Animal Species and Evolution* (1963), he wrote:

> The assumption made by some geneticists, that it was quite impossible to have sensible ideas on evolution until the laws of inheritance had been worked out, is contradicted by the facts. Everyone admits that

Darwin's evolutionary theories were essentially correct and yet his genetic theories were about as wrong as they could be.

Conversely, the early Mendelians, the first biologists (except for Mendel himself) who truly understood genetics, misinterpreted just about every evolutionary phenomenon . . . it is less important for the understanding of evolution to know how genetic variation is manufactured than to know how natural selection deals with it. (Waddington 1957)

Elsewhere in the same book Mayr wrote:

The Mendelian was apt to compare the genetic contents of a population to a bag full of colored beans. Mutation was the exchange of one kind of bean for another. This conceptualization has been referred to as "beanbag genetics". . . genes were believed to be clearly either recessive or dominant; . . . genes were given constant selective values; and there was a tendency to equate genes and characters, as if there were a one-to-one relation." (p. 263)

Mayr further explained how erroneous it is to think in terms of "beanbag genetics":

Work in population and developmental genetics has shown, however, that the thinking of beanbag genetics is in many ways quite misleading. To consider genes as independent units is meaningless from the physiological as well as the evolutionary viewpoint. Genes not only *act* (with respect to certain aspects of the phenotype) but also *interact*. (p. 263)

Viewed against this background, beanbag genetics refers to the early Mendelian concept of treating genes as independent units, and the one gene–one character relationship as well as its application in population genetics to study evolutionary change in mathematical terms. Mayr (1959, 1963) and Conrad Waddington (1957) have called this a "reductionist" approach because, in its simpler version, population genetics ignored any interaction among genes, treating them essentially as independent discrete units.

It is not clear who might be the target of Mayr's criticism. There were clear differences among the three founders of population genetics. Mayr did not specify each target separately—Fisher, Haldane, or Wright. It is well known that epistasis and genic interaction were the foundations of Wright's shifting balance theory. Regarding Haldane, Mayr himself wrote: "Haldane realized that it could lead to deceiving conclusions if one looked at each gene in isolation, because this would fail to reveal synergistic and epistatic interactions." In *The Causes of Evolution*, Haldane (1932a) wrote:

> It is important to realize that the combination of several genes may give a result quite unlike the mere summation of their effects one at a time . . . So selection acting on several characters leads not merely to novelty, but to novelty of a kind unpredictable with our present scientific knowledge, though probably susceptible of a fairly straightforward biochemical explanation. (p. 53)

Mayr's View of Beanbag Genetics

The pioneering contributions of Haldane, Fisher, and Wright were widely recognized and accepted since their beginnings in the 1920s, throughout the 1930s and 1940s, and at least up to 1952, when the first dissonant chord was struck by Waddington (1953), who wrote that the mathematical theory (or theories) developed by Haldane, Fisher, and Wright did not lead to any significant quantitative statements about evolution and that very few qualitatively new ideas had emerged from that body of work. He wrote:

> The formulae involve parameters of selective advantage, effective population size, migration and mutation rates, etc., most of which are still too inaccurately known to enable quantitative predictions to be made or verified. But even when this is not possible, a mathematical treatment

may reveal new types of relation and of process, and thus provide a more flexible theory, capable of explaining phenomena which were previously obscure. It is doubtful how far the mathematical theory of evolution can be said to have done this. (p. 186)

Mayr, too, had entertained similar doubts for several years regarding the true nature of the value of the contributions of Haldane, Fisher, and Wright. Perhaps encouraged by the comments of Waddington, who took the first plunge, Mayr voiced his doubts subsequently in at least three major publications and made a direct attack upon their work in two of them (Mayr 1955, 1959, 1963). There is no evidence to indicate that Waddington's or Mayr's criticisms had any significant impact on the widely held respect and acceptance that had greeted the mathematical theory of evolution from its inception. There were no widespread defections.

In his introductory address to the 1959 Cold Spring Harbor symposium "Genetics and Twentieth Century Darwinism," Mayr summed up what he regarded as the major contribution of the mathematical theory of evolution:

> The main importance of the mathematical theory was that it gave mathematical rigor to qualitative statements long previously made. It was important to realize and to demonstrate mathematically how slight a selective advantage could lead to the spread of a gene in a population. . . . the mathematical theory. . . . in a subtle way it changed the mode of thinking about genetic factors and genetic events in evolution without necessarily making any startlingly novel contributions(p. 3-4).

Earlier in the same address, Mayr recognized that population genetics clearly has two separate roots: One is mathematical and has been widely associated with the names of Fisher, Haldane, and Wright. The other, associated with the names of F.B. Sumner, N.W. Timofeef-Ressovsky, and T. Dobzhansky, is rooted in population systematics and has been generally ignored. His comment that, while the mathematical school has been widely recognized, the other has been ignored, clearly indicates his belief that the importance of mathematical or theoretical population genetics has been grossly exaggerated. Mayr wrote:

> The emphasis in early population genetics was on the frequency of genes and on the control of this frequency by mutation, selection, and random events. Each gene was essentially treated as an independent unit favored or discriminated against by various causal factors. In order to permit mathematical treatment, numerous simplifying assumptions had to be made, such as that of an absolute selective value of a given gene. The great contribution of this period was that it restored the prestige of natural selection, which had been rather low among the geneticists active in the early decades of the century. . . . Yet, this period was one of great oversimplification. Evolutionary change was essentially presented as an input or output of genes, as the adding of certain beans to a beanbag and the withdrawing of others. This period of "beanbag genetics" was a necessary step in the development of our thinking, yet its shortcomings became obvious as a result of the work of the experimental population geneticists, the animal and plant breeders, and the population systematists, which ushered in a third era of evolutionary genetics. . . . These authors [Haldane, Fisher, and Wright] although sometimes disagreeing with each other in detail or emphasis, have worked out an impressive mathematical theory of genetical variation and evolutionary change. But what, precisely, has been the contribution of this mathematical school to evolutionary theory, if I may be permitted to ask such a provocative question. (p.2)

An evaluation of Mayr's statements with reference to mathematical population genetics, especially those concerned with Wright's contributions, clearly indicates that Mayr was very much mistaken. For instance, Mayr criticized that the mathematical population geneticists had only used constant fitnesses in their evolutionary models. Wright's symbol W for selective value has always been defined as applying to a total genotype in the system under consideration, thus involving whatever interaction effects there may be among the component genes. In his contribution to *Haldane and Modern Biology*, (Dronamraju 1968), Wright wrote: "My theory was directed toward ascertaining whether some way, after all, might exist in which selection could take advantage of the enormous number of interactive systems provided by a limited number of unfixed loci."(p.5). The beanbag criticism was particularly inappropriate for Wright, who specifically devised his "shifting balance" theory as a way for a population to go from one harmonious gene combination (most probably "integrated genotype" in Mayr's sense) to another when intermediates were disadvantageous.

Mayr was among those who contributed essays to *Evolution as a Process*, a volume dedicated to Julian Huxley on his 65th birthday (Huxley, Hardy, and Ford 1954). In his essay "Change of Genetic Environment and Evolution," Mayr emphasized that genes do not exist in "splendid isolation," because they are parts of an integrated system. In classical genetics, each locus was studied separately for the sake of simplifying the analysis, treating the genetic factors of an organism as so many beans in a large bag. Mayr wrote: "That this is not so is now known to every geneticist, but 'bean-bag' thinking is still widespread."(p.164) In his address to the 1955 Cold Spring Harbor symposium "Population Genetics," Mayr pointed out that gene interaction is ubiquitous in nature. Hence, evolutionary studies should emphasize the evolution of interactive gene complexes rather than the evolution of single genes. Mayr urged that population geneticists should incorporate a more sophisticated view of the evolutionary process into their mathematical theories of evolution. He emphasized that population geneticists can no longer accept the simple models that sufficed them when they began their research. Mayr wrote:

> Both pleiotropy and polygeny contribute to the genetic cohesion of the population. Pleiotropy contributes because each gene through its manifold effects on development enters into a teamwork with numerous other genes. The replacement of one gene by another in a gene complex may affect numerous interactions of the genotype and lead to an upset of stability which can be compensated only gradually.

Polygeny contributes because selection pressure for or against any aspect of the phenotype will affect all the genes that polygenically add to this character. (1963, pp. 268–269)

Mayr then went on to speculate on the possible impact of the interactions between pleiotropy and polygeny. For instance, are the small phenotypic effects of quantitative inheritance caused by genes whose sole function is to be specific modifiers, or do they represent the pleiotropic effects of genes whose main function lies elsewhere? A major gene whose function is the development of a specific organ, such as the eye, may also serve as a polygene for some other characters such as the body size. The large majority of modifiers are pleiotropic genes with multiple contributions to the phenotype.

Mayr viewed pleiotropy and polygeny as concepts that stress the two end points in the developmental pathway, the gene and the character. He suggested that the continuous interactions among genes all along the epigenetic pathway in numerous ways lead to a strengthening of the "internal cohesion" of the gene pool (1963, p. 269). He wrote a great deal about "genetic cohesion," but we are not told how exactly such a "cohesion" might be attained. It is implied that all these interactions are harmoniously directed at strengthening of the "internal cohesion" of the gene pool.

Epistatic Interactions and Fitness

Mayr wrote a great deal on epistasis and other kinds of genetic interactions. In *Animal Species and Evolution* (1963, p. 270), he considered all interactions between different loci on the same chromosome or on different chromosomes. The picture is further complicated by "penetrance" of the gene, that is, expression, which may be incomplete to varying degrees in different physical or genetic environments.

Another class of interactions are so-called "suppressor" genes, which suppress the effects of other genes; furthermore, they show much variation at each locus in terms of isoalleles of differing strengths.

Because each gene has multiple pleiotropic manifestations that alter the expressed phenotype to varying degrees, and taking into account all the epistatic interactions, Mayr argued that the selective value of a gene is only partially determined by its direct immediate effect on the phenotype. Further complications arise in some species because of the effect of a gene on the fitness enhancing qualities of other genes. Contrary to Mayr's criticism, the founders of population genetics were very much aware of the existence of epistatic interactions and other complications that might affect the selective value of a gene. Indeed, they evaluated the role of such processes in evolution.

It is helpful to understand Mayr's style and approach to such controversial issues. In his classic work on the history of biological thought, Mayr (1982) wrote that he preferred to make decisive and sweeping statements because they would lead to ultimate solutions much more quickly than a middle of the road position. He wrote: "My tactic is to make sweeping categorical statements.... My own feeling is that it leads more quickly to the ultimate solution of scientific problems than a cautious sitting on the fence.... The unambiguous adoption of a definite viewpoint should not be confused with subjectivity" (pp. 9–10). However, a reading of Mayr's writings on the subject makes it clear that he did not appear to have read in depth the original publications of Fisher, Haldane, and Wright and was not thoroughly familiar with their studies on epistasis and interaction.

Mayr's comments are only applicable at the most elementary level of population genetics. For instance, well known population geneticist James F. Crow of the University of Wisconsin wrote:

Indeed elementary expositions of population genetics often fit Mayr's description. But his criticism of the three pioneers was badly misdirected. All of them took linkage, dominance and epistasis into account. One of Fisher's greatest accomplishments was to show that natural selection, despite dominance and epistasis, acts on the additive component of variance, assessed by least squares, and is therefore quite effective. . . . Somewhat later, Kimura showed that under most circumstances, even with linkage disequilibrium, epistatic variance, . . . makes hardly any contribution to the transmissible variance.

Wright's emphasis on epistasis was even greater; complex interactions were the very essence of his shifting balance theory." (2008, p.443)

Crow further wrote:

> Haldane was justifiably critical of Mayr's "genetic cohesion", "integrated gene complexes" and the "coadapted harmony of the gene pool" and one could add "genetic revolution" also to this list. They lack predictive value and add little to our understanding of the basic mechanisms of evolution.
>
> Haldane emphasized that models (or hypotheses, as he preferred to call them) are necessarily simplified, to facilitate testable predictions. Simplification helps to reveal some aspect of science that approximates reality. A poor model, which lacks predictive value, requires improvement. On the other hand, the concepts that Mayr had advanced do not easily lend themselves to model building and predictions.(p.443)

Crow's point that Mayr deeply erred in including Wright in his attack on the mathematical foundations of population genetics was recognized earlier by others. For instance, Provine (1986) wrote:

> Mayr obviously did not appear to understand Wright's great substantively successful efforts to incorporate genic interaction, multiple factors, frequency dependent selection, degree of dominance, and many other factors into his quantitative models, nor did Mayr appear to think that Wright's theory of evolution was anything more than random drift. (p. 482)

Haldane (1964), Crow (2008), and others have characterized Mayr's ideas as vague and untestable. The beanbag model is especially in tune with later developments; in particular, it is a concept that involves random processes that play a major role in molecular evolution. As the models become more and more complex, the bean pool itself may undergo rapid changes, but the random nature of the process remains (Crow 2008).

There is a general impression among population geneticists that Mayr was badly mistaken because either he had not read the original mathematical papers or was unable to understand them. Mayr's criticisms appeared to be directed at an elementary level exposition of population genetics. For instance, Crow (2008) wrote:

> Population genetics is more than Mayr's beanbag genetics. The complications that he said were missing in the model are often taken into account. Population models now include multiple factors, linkage, dominance and epistasis.... In other words, ... population genetics can be regarded as an improved beanbag model. Importantly, it preserves the random property of drawing beans from a bag. (Crow 2008, p.445)

Haldane's Defense

In his defense of beanbag genetics, Haldane (1964) responded to Mayr's criticism. In his address to the 1959 Cold Spring Harbor symposium, Mayr said:

> Fisher, Wright and Haldane have worked out an impressive mathematical theory of genetical variation and evolutionary change. But what, precisely, has been the contribution of this mathematical school to evolutionary theory, if I may be permitted to ask such a provocative question? ... However, I should perhaps leave it to Fisher, Wright and Haldane to point out what they consider their major contributions. (Mayr 1959, p.2)

In his extensive and spirited response to Mayr's criticism, Haldane wrote that his cofounders of population genetics would not respond for different reasons—Fisher was dead and, when he was alive, preferred attack to defense, while Wright was too gentle to respond forcefully. So it was left to him to defend "beanbag genetics." Elementary expositions of population genetics do fit Mayr's characterization. However, Mayr's criticism of the three pioneers, Haldane, Fisher, and Wright, was badly misdirected. It has even been suggested that Mayr had not read the original papers in theoretical population genetics, especially by Wright, who discussed genic interaction quite extensively. Mayr misunderstood the extent of genic interaction that was evident in the contributions of all three pioneers. All three took linkage, dominance, and epistasis into account. Indeed, Fisher (1931) clearly showed that, despite dominance and epistasis, natural selection acts on the additive component of variance, assessed by least squares, and is therefore quite effective. Wright's emphasis on epistasis was far greater than that of either Haldane or Fisher. Complex interactions were central to his shifting balance theory.

Haldane was critical of Mayr's (1963) terminology. He wrote:

> Mayr devotes a good deal of space to such notions as "genetic cohesion," "the coadapted harmony of the gene pool" and so on. These apparently became explicable "once the genetics of integrated gene complexes had replaced the old beanbag genetics. . . . Mayr (1963) attempts to describe this replacement in his chapter 10, on the unity of the genotype. This does not mention Fisher's fundamental paper (1918) on "The correlation between relatives on the supposition of Mendelian inheritance," in which, for example, epistatic interaction between different loci concerned in determining a continuously variable character was discussed. (Haldane1964, p.14)

Haldane went on to point out that Mayr made a large number of enthusiastic statements about the biological advantages of large populations that are unproved and not very probable. Haldane further explained that the genetic structure of a species depends largely on local selective intensities, on the one hand, and migration between different areas, on the other. If there is much dispersal, local races cannot develop; if there is less, there may be clines or races. The "success" of a species can be judged both from its present geographical distribution and numerical frequency and from its assumed capacity for surviving environmental changes and for further evolution.

According to Haldane (1964), there is not enough knowledge to indicate whether a particular species (including our own) would benefit more by increased "cohesion" or gene flow or some other factor from one area to another. Haldane agreed with Mayr (1959, 1963) that beanbag genetics do not explain the physiological interaction of genes and the interaction of genotype and environment. The beanbag geneticist need not know how a particular gene determines resistance of wheat to a particular type of rust, or how it blocks the growth of certain pollen tubes in tobacco, or some other process, in a certain environment. If a beanbag geneticist knows the types of interaction that may occur between a genotype in various environments, he can deduce the evolutionary consequence of these events given further data on various other parameters. Haldane (1964) wrote: "A paleontologist can describe evolution even if he does not know why the skulls of labyrinthodonts got progressively flatter. He is perhaps likely to describe the flattening more objectively if he has no theory as to why it happened."(p.2).

Haldane (1964) cited some examples of his work where mathematical methods yielded deeper insights regarding evolutionary problems that would not have been possible otherwise.[1] In response to Mayr's (1959) critique, Haldane (1964) responded:

> Our mathematics may impress zoologists but do not greatly impress mathematicians. Let me give a simple example. We want to know how the frequency of a gene in a population changes under natural selection. I made the following simplifying assumptions:
>
> (1) The population is infinite, so the frequency in each generation is exactly that calculated, not just somewhere near it,
> (2) Generations are separate. This is true for a minority only of animal and plant species. Thus even in so-called annual plants a few seeds can survive for several years,

---

[1] Quoting David Hume, Haldane (1964) (p.348) commented that algebraic reasoning is not only exact but also imposes an exactness on the verbal postulates made before algebra can be applied.

(3) Mating is at random. In fact, it was not hard to allow for inbreeding once Wright had given a quantitative measure of it,

(4) The gene is completely recessive as regards fitness. Again it is not hard to allow for incomplete dominance. Only two alleles at one locus are considered.

(5) Mendelian segregation is perfect. There is no mutation, non-disjunction, gametic selection, or similar complications.

(6) Selection acts so that the fraction of recessives breeding per dominant is constant from one generation to another. This fraction is the same in the two sexes.

With all these assumptions, we get a fairly simple equation. If q_n is the frequency of the recessive gene, and a fraction k of recessives is killed off when the corresponding dominants survive, then

$$q_{n+1} = \frac{q_n - kq_n^2}{1 - kq_n^2}$$

Haldane (1964) acknowledged that the mathematician H.T.J. Norton had given an equivalent equation in 1910, and Haldane (1924) produced a rough solution when selection is slow, that is when k is small. However, he pointed out that even such a simple-looking equation would not yield a simple relation between q and n. Many years later, Haldane and Jayakar (1963) solved this equation in terms of automorphic functions. Haldane noted that the mathematics are not much worse when inbreeding and incomplete dominance are taken into account. But they are much more complicated when selection varies from year to year and from place to place or when its intensity changes gradually with time. When such problems are solved, the mathematics employed would be truly impressive.

Haldane's point was that the mathematical methods used by himself and other founders of population genetics, which impressed Mayr so much, are in fact not impressive at all from a mathematician's point of view. And the few mathematicians who have interested themselves in such matters have been quite unhelpful.

Examples from Previous Work   Haldane cited examples from his work showing that verbal arguments are liable to be fallacious. For instance, in the course of building his mathematical theory of natural selection, Haldane calculated the equilibria between mutation of various types of genes and selection against them. He was then able to estimate human mutation rates and was the first to do so (Haldane 1932a). The estimation of human mutation rates was a by-product of Haldane's mathematical theory of natural selection. His method was used by others later to estimate mutation rates of various kinds. However, certain questions remain unanswered. Selection and mutation must balance in the long run, but how long is long enough? In two mathematical papers, Haldane (1939, 1940) showed that while harmful dominants and sex-linked recessives reach equilibrium "fairly quickly," the time needed for the frequency of an autosomal recessive to get halfway to equilibrium after a change in the mutation rate, the selective disadvantage, or the mating system may be several thousand generations. Haldane wrote that the verbal argument is liable to be fallacious in such cases. "As few people have read my papers on the spread or diminution of autosomal recessives, and still fewer understood them, the 'balance' method, which I invented, is applied to situations where I claim that it leads to false conclusions" (Haldane 1964, p.347).

Another example concerned industrial melanism in the coal mining regions of industrial England. From the records of the spread of the autosomal gene for melanism in the peppered moth, *Biston betularia*, in English industrial districts, Haldane (1924) calculated that it conferred a selective advantage of about 50 percent on its carriers. However, as Haldane pointed out in his "defense," few or no biologists accepted his conclusion at that time. Those biologists who gave any thought at all to quantitative thinking were accustomed to thinking of the order of 1 percent or less. However, many years later, Kettlewell's (1956) research showed that in one wood the melanics had at least double the fitness of the original type. This was the first example of the measurement of intensity of selection in nature. Haldane (1964) commented: "If biologists had had a little more respect for algebra and arithmetic, they would have accepted the existence of such intense selection thirty years before they actually did so"(p.348). Haldane further (1964) wrote: "The mathematics on which my conclusion was based are not difficult, but they are clearly beyond the grasp of some biologists."(p.349)

Another example concerned selection against hemolytic disease in newborns. When Landsteiner and Wiener discovered the genetic basis of human fetal erythroblastosis, Haldane (1942) pointed out that the death of Rh-positive babies born to Rh-negative mothers could not lead to a stable equilibrium, suggesting that the modern human populations of Europe were the result of crossing between original Europeans, most of whom possessed Rh-positive genes, and other peoples who had a majority of Rh-negative genes. Haldane wrote that a distinguished colleague had calculated an equilibrium but had not dipped far enough into the bag to notice that it was unstable. Later, two relict populations were discovered, in northern Spain and in Switzerland, with a majority of Rh-negative genes.

More Exposition Instead of Mathematics? Haldane (1964) in his famous "defense" indicated that Fisher, Wright, and himself should have spent more time in an exposition of their methods and applications. He wrote: "Perhaps a future historian may write, 'If Fisher, Wright, Kimura, and Haldane had devoted more energy to exposition and less to algebraical acrobatics, American, British and Japanese genetics would not have been eclipsed by those of Cambodia and Nigeria about A.D. 2000.' I have tried in this essay to ward off such a verdict."(p.358) Among the trio of the founders of population genetics, Haldane was the only one who provided nonmathematical expositions from time to time. Fisher and Wright were totally silent in this respect.

Mendelian Approach

The concepts underlying the "beanbag" controversy can be traced back to the fundamental differences between the Darwinian and Mendelian approaches to biology. The mathematical theory of evolution developed by Haldane, Fisher, and Wright followed the reductionist conceptualization that was employed by Gregor Mendel (1866) himself in his pioneering studies of transmission of characters. As Carlson (2004) aptly put it:

> If I were to credit him for a particular discovery or contribution to science, I would single out his conception of breeding analysis. Mendel's scientific outlook was that of an ultimate reductionist. Where his predecessors looked for holistic or cosmic significance to the meaning of hybrids, Mendel decided to strip hybrids of their mystical appeal and tease apart the components of the two parents in the hybrids by focusing on one trait at a time. He also used an approach common to some of the most successful scientists, for example, Thomas Hunt Morgan (who used fruitflies) or Seymour Benzer (who used bacteriophages), he swept all the difficult, complex and messy problems under a mental rug and started his research with the simplest components available to him. (p. 47)

We should also remember that prior to the work of Haldane, Fisher, and Wright, Hardy (1908) and Weinberg (1908) derived their fundamental equation of population genetics, and the method they followed was definitely of the "beanbag genetics" variety. Indeed, their work has been a pillar of "beanbag genetics." Subsequently, Morgan et al (1915), in their experiments with the fruitfly *Drosophila*, pioneered gene mapping, once again using the methods and concepts of beanbag genetics.

Darwinian Approach

In contrast to Mendelian genetics, the very nature of the larger questions that Charles Darwin was attempting to answer at the population and species levels required a broader approach. As Mayr (1959, 1963, 2004) reminded us repeatedly, Darwin introduced population thinking. Mayr (2001) wrote:

> Darwin made a radical break with the typological tradition of essentialism by initiating an entirely new way of thinking. . . . What we find among living organisms, he said, are not constant classes (types), but variable populations. . . . Darwin's new way of thinking, being based on the study of populations, is now referred to as population thinking. . . . The gradual replacement of essentialism by population thinking led to long-lasting controversies in evolutionary biology. . . . It is the foundation of modern evolutionary theory." (From *What Evolution Is*,(Mayr (2001), p. 75)

Mayr continues: "Population thinking is one of the most important concepts in biology. It is the foundation of modern evolutionary theory and one of the basic constituents of the philosophy of biology."(Mayr (2001) p.75)

Darwin, in his approach to population thinking, recognized two aspects of evolution. One is the "upward" movement of a phyletic lineage: its gradual change from an ancestral to a derived condition (*anagenesis*). The second consists of the splitting of evolutionary lineages or the origin of new branches of the phylogenetic tree. The second aspect involves the creation of biodiversity (*cladogenesis*), and it always begins with an event of speciation. Mayr emphasized both aspects.

Correspondence between Haldane and Mayr

The correspondence between Haldane and Mayr contains several references to beanbag genetics, each reiterating his own position on the subject from different perspectives, a minuet of sorts that is interesting to watch. In his letter to Mayr, dated May 8, 1963, Haldane wrote:

> I am an unrepentant "beanbag geneticist." Non-mathematicians often fail to realize the complexity of behavior, and sometimes the

self-regulatory capacity, of material systems composed of simple components. Newton thought the creator put the beans (sun, planets and
satellites) in the bag and given it a shake. But he thought the system
would lose its regularity, and after a few thousand years the creator
would have to give it another push. . . . I have got back to beanbag
genetics in a big way . . . there are a whole lot of conditions other than
superiority of heterozygotes which will conserve polymorphism on a
reasonable scale. 30 years ago I showed that mutation would not, unless
selective differentials were as small as mutation rates, but that migration might do so. We can now give the conditions as to migration
rather more concretely.

In the following paragraph, Haldane informed Mayr that he was
investigating a situation (with S.D. Jayakar) when an initially "unfavoured" genotype gradually increases its selective value. The population may change rather suddenly, even if the relative fitness is only
increasing slowly, and the selection of "modifiers" may make this
change still more sudden. (In the margin of the letter at this point
Mayr wrote "no longer beanbag!") Haldane was referring to what
might happen when climatic conditions change slowly.

Later, in the same letter, Haldane wrote:

We have just worked out a fairly comprehensive theory of what
happens under selection of constant intensity when this is fairly strong
(as it doubtless is when a new niche is occupied and there is no immigration from the old one). For this we need automorphic functions of
a kind which were fashionable in France about 1920. I may of course
be hopelessly out-of date in my approach. I am sure bright boys like
Jim Crow think so. But it seems to me that mathematical genetics are
still about the stage of $2 s = 1/2$ ft, and that the mathematicians who
come in from time to time are interested in inessentials, or shall we say,
topics whose biological importance is not obvious.

One year before the publication of his famous "defense" of beanbag genetics Haldane informed Mayr of its impending publication. In
his letter dated June 3, 1963, Haldane wrote:

> I have also completed a much more serious attack on you, entitled "A defense of beanbag genetics." This is intended for "Perspectives in biology and medicine." I may say that from defense I pass to counter-attack. However, I am going to get it typed, and then look at it again after three months or so, to see if I find any sections of it unclear or unfair, or whether perhaps some new arguments have occurred to me.

He goes on to mention an example of his "counter-attack" involving strong selection for the carbonaria mutant of *Biston betularia*, which Haldane calculated in 1924. He wrote: "This was beanbag genetics, and nobody took it seriously for 30 years." In the same letter, Haldane referred to his first paper on the mathematical theory of natural selection (Haldane 1924).

In his letter to Haldane, dated June 18, 1963, Mayr responded to Haldane's warning:

> I shall receive your "attack" on me philosophically. In a big volume like the one I have written, it is quite impossible to avoid shortcuts and generalizations. For instance, I had your 1924 paper in an earlier draft, but took it out since you refer to it in your later papers, and I had to streamline my over-long bibliography. I wonder whether other readers would also come to the conclusion that I "suggest that you did not believe in strong selection until 1957." The whole point I tried to make was that around 1930, the emphasis was on the effect of slight differences in selection pressure and that this led many non-geneticists into making unrealistic assumptions.

One year later, in a letter dated June 3, 1964, Mayr wrote to Haldane that the editor of *Perspectives in Biology and Medicine* had asked him to respond to Haldane's defense of beanbag genetics. However, Mayr declined to do so and stated in his letter to Haldane:

> Obviously most of what you say will be fully endorsed by me also. It is all a matter of emphasis. A certain amount of beanbag genetics is the necessary basis for all else, but on the whole, beanbag genetics is

singularly unsuitable to explain any but the most elementary evolutionary phenomena. Penetrance, as you rightly remark in your letter, is a case in point. The recent models developed by Jacob and Monod, and others, about regulating genes are further substantiating arguments. The enormous increase in the amount of DNA among the eucaryotes, most of it apparently not used for structural but for regulating genes, is still further evidence.—I still believe that beanbag genetics in many cases had a detrimental rather than beneficial influence on evolutionary thinking. You, yourself, without using such terminology, have called attention to this in many of your papers.

In the same letter, Mayr referred to the delayed election of Haldane to the U.S. National Academy of Sciences, stating that

your election to the National Academy is not an honor to you but a removal of a disgrace from us. We have an excessively democratic system of voting in ordinary members, but Foreign members are proposed by the Council, and this means that the influence of the biologists is rather limited. I do not know whether this will please you or not, but I can assure you that you have been for years the leading candidate of the biologists. It has been a great source of satisfaction for all of us that your election was finally ratified. Hence even if the election should not mean a thing to you, I can assure you that it means a lot of us.

The Species Concept

Mayr's Definition of a Species   In his spirited defense of Darwin and his theory of evolution, Mayr (2001, pp.163-165) wrote that naturalists have had a terrible time trying to reach a consensus on what species are. In various writings, this has been known as "the species problem." Mayr (2001) wrote:

Even at present there is not yet unanimity on the definition of the species. This is in part due to the fact that the term "species" has been applied to two very different things, to the species as concept and to the species as taxon. Another complication has been that most naturalists have changed from an adherence to the typological species concept to an acceptance of the biological species concept in the last 100 years (p. 163–165).

In *Animal Species and Evolution* (1963), Mayr described the biological species concept in detail: Traditional taxonomic descriptions of species emphasized differences in morphological characteristics that are most useful diagnostic purposes. However, species may also differ from each other in size, color, internal structure, physiological characters, cell structures, chemical constituents (particularly proteins and nucleic acids), ecological requirements, and behavior. Mayr further added that every species is the product of a long history of selection and is thus well adapted to the environment in which it lives. Emphasis has been laid by both Mayr (1940, 1963) and Dobzhansky (1950) that a species represents the largest and most inclusive reproductive community of interbreeding individuals that share a common gene pool and reproductively isolated from other such groups.

Three important aspects of the biological species concept were highlighted by Mayr (1957): (a) Species are defined by distinctness rather than by difference; (b) species consist of populations rather than of unconnected individuals; and (c) species are more unequivocally defined by their relation to nonconspecific populations ("isolation") than by the relation of conspecific individuals to each other. The decisive criterion is the reproductive isolation of populations.

Haldane had already foreshadowed what was later called the "biological species concept." In *The Causes of Evolution* (1932), Haldane wrote: "It is unfortunately impossible to give a satisfactory definition of the term 'species'... In many cases the species may be defined as a group of organisms which can breed together without loss of fertility in the first or subsequent generations." (p. 35–36) Later, Haldane (1954) presented a species as something in equilibrium under fairly intense "forces."

Mayr persuaded Harvard University Press to send the proofs of his book *Animal Species and Evolution* to Haldane in India. In response, on April 6, 1963 Haldane wrote: "Thank you for the proofs, which I am reading. Whether people agree with your conclusions or not your book will be an invaluable guide to the literature." However, Haldane disagreed with Mayr's view of a species. He wrote:

> "*As a bean-bag geneticist*"[2] I think your view of a species may be a little too physiological and not historical enough. No doubt the various genes (as regards frequency and location as well as molecular pattern) have to fit together to form, if not an adaptive peak in Wright's sense, range of such peaks, separated at most times from other such ranges by deep "valleys." But on reading you I sometimes got the feeling that you think we could calculate the species if we knew enough about the genes.

Haldane added that existing species are only a small fraction of those that might have been made up with the genes available in a genus or family, and that the reasons for their existence are largely historical. Haldane further characterized Mayr's definition of a species as being too "futuristic" for his taste, although he approved Mayr's description of a species.

Target of Selection   At the center of the beanbag controversy is the target of natural selection. Haldane extended the mathematician

---

[2] Wilhelm Johannsen (1857–1927) introduced the terms "genotype" and "phenotype." His work did not focus on individual genes. He was studying a quantitative trait: bean size. His method was inbreeding and selection. He selected for the extremes—smallest and the largest size possible. He showed that he could select for a given size (small or large) for about ten or so generations, after which selection no longer worked. He then claimed he formed a pure line and that its genotype was identical for all members of that line. But since the members of that small bean group varied, he claimed they differed in phenotype from each other because of environmental events.

In 1922, Haldane was in contact with Norton, who was at Trinity College, Cambridge, and had prepared the selection table in R.C. Punnett's (1915) book *Mimicry in Butterflies*. As a student of the famous mathematician G.H. Hardy, Norton had excellent credentials. Because of illness and other factors, Norton did not publish his full results until 1928. By then, Haldane had published five papers on the mathematical theory of natural selection, while acknowledging that Norton's table motivated him to pursue his own mathematical investigations.

Fisher and Haldane used individual genes as targets of selection, calculating evolutionary change in terms of changes in gene frequencies. Mayr (1988) commented:

> Accepting the change of gene frequencies as the earmark of evolution, some authors unfortunately began to ignore whether such change was due to genetic drift or to selection. And in the single-minded concentration on genes, it was often forgotten how important individuals, populations, and species are in determining the course of evolution. (pp. 100–103).

In the nineteenth century as well as during the years following the rediscovery of Mendel's laws in 1900, the distinction between phenotype and genotype was not known. The organism as a whole was considered to be the target of selection. However, that view had changed with the discoveries of W. Bateson and W. Johannsen.

Bateson (1861–1926) recognized phenotypic variation about the same time as Johannsen (1857–1927) and that a character trait could be complex. Bateson used plant breeding to obtain modifications of 9:3:3:1 ratios, and he interpreted 15:1, 9:7, or 9:4:3 ratios involving phenotypic classes that had different genotypes resulting in a common phenotype. Bateson (1906) introduced the terms "genetics," "allelomorphs," and "coupling" and "repulsion" in linkage, which have played a pioneering role in developing the genotype/phenotype concept in the years preceding the discoveries of the Morgan school. The term "gene" itself was suggested by Johannsen in 1909 to replace the term "factor" which was used by Bateson. Many years later, in 1942, Haldane introduced the terms "cis" and "trans" into genetics, replacing "coupling" and "repulsion." Those were the years when the genotype/phenotype relationship, as we know it today, emerged.

Thomas Hunt Morgan (1866–1945) and his students adopted Johannsen's view and applied it to the expression of individual genes. Muller showed that there were modifier genes and claimed that Darwinian natural selection utilized the phenotype to select for genotypes involving chief genes and genetic modifiers. He used this to explain the evolution of dominance of a dominant trait. It must be emphasized, however, that the first person to recognize that a genotype is not the same as a phenotype was the founder of genetics, Mendel himself! He drew the distinction between the peas heterozygous for yellow (i.e., Yy) and the peas homozygous for yellow (i.e., YY), which are seen as the same phenotype (Y). That simple distinction between phenotype and genotype by Mendel is very different from Johansson's approach as well as Morgan's group, which saw in the continuum of expression a way to expand Darwinian natural selection and account for the origin of gradual change over time. With the establishment of the gene as the basic unit of inheritance, Fisher and Haldane based their mathematical calculations of evolutionary change in terms of gene frequencies.

Individual as the Target of Selection   Mayr (1988) preferred the individual as a whole rather than each separate gene as the target of selection because he considered the genotype to be a well-integrated system analogous to an organism with structure and organs. Other reasons cited by Mayr include the following. First, it is the individual as a whole that either does or does not have reproductive success. Second, the selective value of a particular gene may vary greatly depending on the genotypic background. Third, since different individuals of the same population differ at many loci, it would be exceedingly difficult to calculate the contributions of each of these loci to the fitness of a given individual. Fourth, accepting the individual as a whole makes it unnecessary to make the confusing distinction between internal selection and external selection. Finally, the individual as the target of selection implies the individual at all stages of its life cycle, from the fertilized egg through all stages up to death.

Unit of Selection   Is the unit of selection an apt term? Mayr referred to the *selection of an object* by the term "target of selection." Mayr found it gratifying that Elliott Sober (1984) agreed with his own view that to characterize the gene as the unit of replication establishes very little about what the unit of selection is. Both were in agreement in rejecting the claim of the "genic selectionists" that the unit of selection must have a high degree of permanence, and that only the gene, the unit of replication, qualifies. Mayr wrote that variation, rather than permanence, is the principal prerequisite for selection. Mayr criticized that the mathematical population geneticists had only used constant fitnesses in their evolutionary models.

In contrast to Mayr's characterization, one of the founders of population genetics, Sewall Wright has always defined his symbol W for selective value as applying to a *total* genotype in the system under consideration, thus involving whatever interaction effects there may be among the component genes (Wright 1960). In his 1968 contribution to *Haldane and Modern Biology*, Wright wrote: "My theory was directed toward ascertaining whether some way, after all, might exist in which selection could take advantage of the enormous number of interactive systems provided by a limited number of unfixed loci." (p. 5)

Neglected Aspect of Selection   Many years later, Mayr (1959) acknowledged that Haldane had, in fact, drawn attention to relative selective values and the fact that selection acts on phenotypes, not genes. He was referring to Haldane's paper on the cost of natural selection (1957b). Mayr commented that Haldane had called attention to a neglected aspect of selection: "How severe a selection pressure can a population endure at any one time?" In fact, Haldane (1957b) began his paper on the cost estimate with the following sentences:

> It is well known that breeders find difficulty in selecting simultaneously for all the qualities desired in a stock of animals or plants. This is partly due to the fact that it may be impossible to secure the desired phenotype with the genes available. But, in addition, especially in slowly breeding animals such as cattle, one cannot cull even half the females, even though only one in a hundred of them combines the various qualities desired. (p. 511)

Haldane (1957b) noted that the situation with respect to natural selection is comparable. He quoted Kermack (1954), who showed that characters that are positively correlated in time may be negatively correlated at any particular horizon: "Genes available do not allow the production of organisms which are advanced in respect of both characters." (p.511). Haldane (1957b) stated that he was attempting to quantify the statement that natural selection cannot occur with great intensity for a number of characters at once unless they happen to be controlled by the same genes. Indeed, Haldane (1957b) made it clear that he was thinking of both the statics and the dynamics of evolution and that the selective value of a gene may vary depending on several factors, such as its interaction in different backgrounds and environments, among others.

Two important factors were emphasized by Haldane (1957b). First, the loss of fitness in genotypes and its impact on the population depend on the stage of the life cycle at which it occurs and the ecology of the species concerned. In some species, the failure of a few eggs or seeds to develop will have little impact on the ultimate increase in its numbers. Examples are the work of Durham (1911) on prenatal mortality in mice and the work of Salisbury (1942) on the effect of elimination of a fraction of seeds in several plant species. On the other hand, death or sterility at a later stage is more serious in species that compete with another for food, space, light, and so on.

Second, Haldane (1957) pointed out that in those parts of its habitat where climate, food, and other factors are optimal, the density of a species is usually controlled by negative density-dependent factors, such as disease, especially infectious disease, promoted by overcrowding, competition for food and space, among others. Haldane had discussed this in greater detail in his 1949 paper "Disease and Evolution." In some areas, a moderate fall in fitness has little effect on the density, but in the parts of the habitat where the population is mainly regulated by density-independent factors, such as temperature and salinity, the species can only maintain its numbers by increasing its fitness to its fullest extent. In marginal areas, the species may even disappear altogether (Birch 1954).

Evolution as a Process In "The Statics of Evolution," Haldane (1954) wrote that he tried to present a species as something in equilibrium under fairly intense "forces." For instance, the struggle between selection for fertility and longevity may be an intense process, because different genotypes of nearly equal net fitness differ greatly in these respects. Paleontological evidence indicates that over long periods the equilibrium may be altered very gradually.

Haldane discussed several situations that may favor rapid selection. For instance, evolution by natural selection can be very rapid "if a species, like the first land vertebrates, or the first colonists of an island, finds itself in an environment to which it is very ill-adapted, but in which it has no competition, and perhaps no predators and few parasites" .(p-513-514). Haldane further noted that selection might be so intense as to reduce the capacity for increase to one-tenth of that of its adapted descendants, and it could yet hold its own. He added that the rate of evolution is set by the number of loci in a genome, and the number of stages through which they can mutate. For instance, if pre-Cambrian organisms had much fewer loci than their descendants, they may have evolved much quicker, although with limited possibilities.

Wright's Views

Wright (1931) stated that the early geneticists were interested in characters mainly as markers for genes. For the most part, they were interested only in "good" genes that were consistently associated with easily classifiable characters. However, interaction effects were recognized very early (Cuenot 1904; Bateson 1909) and interpreted as due to epigenetic chains of reactions. Natural selection acts directly upon individuals and their genotypes as a whole, not on individual allelic genes. The most important "character" in this connection is the "selective value" of the genotype.

Some Fundamental Principles Enunciated by Wright

The variations of most characters are affected by a great many loci (the multiple factor hypothesis).

Generally speaking, each gene replacement affects many characters (universal pleiotropy).

Each of the possible alleles at any locus has a unique array of differential effects on taking account of pleiotropy (uniqueness of alleles).

The dominance relation of two alleles is an attribute not of them but of the whole genome and of the environment. Dominance may differ for each pleiotropic effect and is generally modifiable. The effects of multiple loci on a character in general involve much nonadditive interaction.

Both ontogenetic homology and phylogenetic homology depend on calling into play similar chains of gene-controlled reactions under similar developmental conditions.

The contributions of measurable characters to overall selective value usually involve interaction effects of the most extreme sort because of the usually intermediate position of the optimum grade, a situation that implies the existence of innumerable different selective peaks (multiple selective peaks).

Contrary to Mayr's characterization, Wright discussed genic interactions extensively in his papers in population genetics. He wrote that interaction effects are least likely where the observed character is closely related to primary gene action as it is in the case of allelic differences in protein composition. He then considered the serologic responses of antigens because "they are not so close, but are also largely independent of the rest of the genome." Wright (1968b, p. 71) noted that Irwin (1947) had observed that the hybrids of many species of doves and of certain ducks exhibit "hybrid" antigens not present in either parent, and that Fox (1949) demonstrated interactions among the antigenic effects of mutations induced in *Drosophila melanogaster* by X-rays.

Wright (1968b) further observed that interaction effects necessarily occur in relation to the ultimate products of chains of metabolic processes in which each step is controlled by a different locus. It is implied that interaction effects are universal in the more complex characters that trace to such processes. They are usually accompanied by effects on dominance as well. Wright's observations were based on his breeding experiments to study the genetics of coat color pigments in guinea pigs.

In his classic paper of 1931, Wright wrote:

> If the population is not too large, the effects of random sampling of gametes in each generation bring about a random drifting of the gene frequencies about their mean positions of equilibrium. In such a population we can not speak of single equilibrium values but of probability arrays for each gene, even under constant external conditions. If the population is too small, this random drifting about leads inevitably to fixation of one or the other allelomorph, loss of variance and degeneration.(p. 97).

Wright went on to explain how conditions might be more favorable in a population of certain intermediate size for the appearance of new adaptive combination of types, leading eventually to more favorable conditions favoring an evolutionary process than in the scheme proposed by Fisher (1930):

> At a certain intermediate size of population, however (relative to prevailing mutation and selection rates), there will be a continuous

kaleidoscopic shifting of the prevailing gene conditions, not adaptive itself, but providing an opportunity for the occasional appearance of new adaptive combinations of types which would be never reached by a direct selection process. There would follow thoroughgoing changes in the system of selection coefficients, changes in the probability arrays themselves of the various genes and in the long run an essentially irreversible adaptive advance of the species. It has seemed to me that the conditions for evolution would be far more favorable here than in the indefinitely large population of Dr. Fisher's scheme. . . . A much more favorable condition would be that of a large population, broken up into imperfectly isolated local strains. (p.97)

Wright then emphasized that his scheme is not limited by the mutation rate:

The rate of evolutionary change depends primarily on the balance between the effective size of population in the local strain and the amount of interchange of individuals with the species as a whole and is therefore not limited by mutation rates. The consequence would seem to be a rapid differentiation of local strains, in itself nonadaptive, but permitting selective increase or decrease of the numbers in different strains and thus leading to relatively rapid adaptive advance of the species as a whole. (p.97)

Intermediate Optima and Multiple Selective Peaks　　With respect to selection, Wright (1931) wrote that selection, whether in mortality, mating, or fecundity, applies to the organism as a whole. A gene that is more favorable than its allele in one combination may be less favorable in another. Even when the effects are cumulative, there is generally an optimum grade of development of the character, and a given gene may be favorably selected in combinations below this optimum but selected against in combinations above the optimum. This situation can be illustrated by a series of guinea pig colors in which replacements mainly act cumulatively. Interaction between various genes results in different selective peaks.

In particular, Wright (1931) considered multiple selective peaks in relation to pleiotropy. He wrote:

> The prevalence of pleiotropy also tends to bring about multiple selective peaks. Genes that have favorable effects at all will, in general, also have one or more unfavorable effects, with the net effect depending on the total array of genes. Evolution depends on the fitting together of favorable complexes from genes that cannot be described in themselves as either favorable or unfavorable. (p.97)

Wright wrote further that the simplest model for dealing with selective value as a character seems to be the cumulative action of many genes on an underlying character, but with selective value falling away from an intermediate optimum, with the multiple resulting selective peaks differentiated by second-order pleiotropic effects. Generally speaking, the existence of complex patterns of factor interactions must be taken as a major premise in any serious discussion of population genetics and evolution.

Wright's well-known idea was that the stochastic factors might play an important role allowing a population to explore the adaptive landscape. In his "shifting balance" formulation, the division of an abundant species into many small subpopulations tends to maximize the species' ability to evolve toward higher fitness peaks, because a small group might have a fortuitous combination of alleles allowing it to move to a higher fitness peak. This model has been controversial even up to the present day, because of our lack of knowledge about the characteristics of "fitness landscapes."

Is Beanbag Genetics Obsolete?

Far from being obsolete, Haldane argued that "beanbag" genetics had hardly begun its triumphant career. It needs an arsenal of mathematical tools as well as accurate numerical data. Haldane stated that one of the important functions of beanbag genetics is to show what kind of numerical data are needed.

With reference to physiological genetics, Haldane in his "defense" responded by saying that the mathematical theory of physiological genetics is about fifty years behind that of beanbag genetics. If a metabolic process depends on four enzymes acting on the same substrate in succession, one can calculate what will happen if the amount of one of them is halved, provided that one is working with enzymes in solution in a bottle. However, Haldane stated that we know far too little of the structural organization of living cells at the molecular level to predict what will happen if the amount (of an enzyme) is halved in a cell, as it is in some heterozygotes. If the enzyme molecules are arranged in organelles containing just one of each kind, the rate of the metabolic process will probably be halved. But if they are in a random or a more complicated arrangement, it may be diminished to a slight extent, or even increased, for the activities of some enzymes are inhibited by an excess of their substrate. As Haldane noted, this may rarely lead to heterosis.

Haldane offered his own counteroffensive in response to Mayr's comments in *Animal Species and Evolution* (1963).

First, Haldane pointed out that when Mayr discussed how sympatric speciation might occur, his arguments are always verbal rather than algebraic. And occasionally they are inconsistent. Referring to Mayr's (1963) statements, Haldane wrote:

> On page 473 he makes seven assumptions, of which (1) is "Let A live only on plant species 1," and (4) is "Let A be ill adapted to plant species 2. These two assumptions seem to me to be almost contradictory. If A lives only on species 1, the fact that it is ill adapted to species 2 is irrelevant. If emus live only live in Australia, the fact that they are ill adapted to the Antarctic has no influence on their evolution. If the assumptions had been "(1) Let A females only lay eggs on species 1," and "(4) Let A larvae (not all produced by A mothers) be ill adapted to species 2," I could have applied mathematical analysis to the resulting model. I propose to do so in the next few years. But I hope I have given enough examples to justify my complete mistrust of verbal arguments where algebraic arguments are possible, and my skepticism

when not enough facts are known to permit of algebraic arguments. (Haldane 1964, p.13)

Second, Haldane further pointed out that in earlier chapters of his book, Mayr (1963) appeared to have been not closely familiar with early literature of beanbag genetics:"On page 215 (quoted by Haldane) Mayr wrote:'the classical theory of genetics took it for granted that superior mutations would be incorporated into the genotype of the species while the inferior ones would be eliminated.'" Haldane commented that the earliest post-Mendelian geneticists, such as Bateson and Carl Correns, wrote very little about this matter. Fisher (1922) pointed out that if a heterozygote for two alleles was fitter than either homozygote, neither allele would be eliminated. Haldane mentioned that Fisher's finding had been a well-established conclusion of beanbag genetics at least since 1922. In the second paper of his series, "Mathematical Theory of Natural Selection," Haldane (1926) referred to Fisher's conclusion and extended it slightly.

Third, Haldane referred to Mayr's terminology such as "genetic cohesion" and "the coadapted harmony of the gene pool" and commented that many such statements in chapter 10 of Mayr's book, especially on the biological advantages of large populations, are unproven and not very probable. It is ironic, Haldane wrote, that some of Mayr's (1963) statements can only be proved by using the methods of beanbag genetics, but the mathematics needed will be "exceedingly stiff."

Fourth, referring to Mayr's idea of genetic "cohesion," Haldane wrote that we do not have enough knowledge in any species (including our own) to say whether it would be benefited by more or less "cohesion" or gene flow from one area to another. Haldane wrote: "The 'success' of a species can be judged both from its present geographical distribution and numerical frequency and from its assumed capacity for surviving environmental changes and for further evolution."(Haldane 1964, p.15)

Fifth, Haldane mentioned the specific example of intercaste marriages in India. These are increasing in numbers today because of increased travel away from home for education and employment.

Traditional family influences are being disrupted under the pressure of a rapidly and upwardly mobile society. This situation is a radical departure from centuries of tradition where marriage outside the caste was unthinkable. Intercaste marriages are disturbing long-established gene pools. As Haldane pointed out, if intercaste marriages become common, "various undesirable recessive characters will become rarer; but so may some desirable ones, and the frequency of the undesirable recessive genes, though not of the homozygous genotypes, will increase."(Haldane 1964, p.15)

Finally, Haldane added that since extinction is the usual fate of every species, even if it has evolved into one or more new species, Mayr's optimism of chapter 10 does not seem justified.

Beanbag Genetics Today

On the current position of "beanbag genetics," Crow (2008, 2009) has written eloquently. I am indebted to James Crow for the following points. Beanbag genetics has kept up with later developments in one respect. Drawing colored beans from a bag forces one to attend to random processes, which play a vital role in recent theories of molecular evolution. Although the "bean pool" may change with time, the essential randomness of the process remains. The essential nature of "beanbag" genetics will be maintained.

Developmental genetics increasingly feeds on the advances made in population genetics, and vice versa. One example is quantitative trait locus mapping. The techniques are those of transmission genetics, but the underlying mechanism is based on understanding the developmental basis of quantitative traits. This interaction will continue and expand in the future.

Today, gene action is much better understood. The important role of regulation with the continuing discovery of new mechanisms has replaced the earlier emphasis on transcription and translation. Simultaneously, as our understanding of physiological genetics has improved, so has the importance of beanbag genetics. The scope of beanbag genetics has greatly increased because molecular techniques have provided an abundance of new data that was lacking previously. Whereas in classical genetics gene was the object of main focus, today attention has shifted to nucleotides.

Beanbag genetics today includes molecular clocks, nucleotide diversity, coalescence, and DNA-based phylogenetic trees, as well as the four major holdovers from the classical period: mutation, selection, migration, and random genetic drift. Rates of evolution at the nucleotide level can be measured and compared among diverse populations and among species. But most important, genetic differences between populations could be measured only by hybridization. Many species differences, not to mention genera and higher orders, could not be measured because the groups were not crossable. Today it is commonplace to compare DNA differences and similarities in diverse species, orders, and taxa. Finally, the most important evidence for the concept of "out of Africa" in human ancestry came from nucleotide diversity—in other words, beanbag genetics. (Crow 2009)

Haldane, Fisher, and Wright, though differing among themselves, have developed a fairly cohesive theory that places natural selection as the main guiding force in evolution. The new beanbag genetics includes Motoo Kimura's neutral theory. Molecular studies have revealed the striking fact that an overwhelming fraction of DNA is noncoding. Much of the evolution in these regions is driven by neutral mutations, as Kimura argued, and the fate of individual mutants appears to be determined by random drift. These concepts became the foundation for the idea of molecular clocks as well as the idea of null hypothesis for measuring selection. These developments have confirmed further the prediction of Haldane that beanbag genetics has hardly begun its triumphant career (Morton 2008, Singh 2003).

The rise of molecular methods has led to an increase in the importance of mathematics in population genetics and evolution. The abundance of data that require mathematical analysis has greatly increased. At the time of Mayr's challenge, evolution had a beautiful theory but very few opportunities to apply it. Now the situation is reversed: Data appear faster than existing theory can deal with them.

That mathematics will play an increasingly important evolutionary role in the near future seems clear.

Recent mathematical work has gone well beyond that of the three pioneers. This is due partly to skilled mathematicians entering the field and bringing new techniques with them; especially noteworthy are stochastic processes. But perhaps more important is the extensive use of computers—often you can use a computer to get by without deep mathematical knowledge. An additional influence is the explosive growth of molecular data, which lend themselves to mathematical treatment. In the first half of the twentieth century, population genetics and evolution had a beautiful theory, but there were very limited opportunities to apply it. Now the situation is reversed: Molecular data accumulate too fast to be assimilated.

Phylogenies Mathematics, especially computers, have become indispensable to Mayr's own field, systematics. A mammalian DNA sequence supplies billions of bits of information, thus for the first time providing an opportunity for a procedure independent of personal judgments. Molecular analysis of DNA sequences, using the new methods, has shown that our closest relatives are chimpanzees. Surprisingly, we share some 90 percent of our DNA with mice, dogs, and elephants. This is no surprise to those acquainted with the neutral theory. These numbers are fully consistent with expectations based on mutation rates and the times involved. New computer programs facilitate working out phylogenies and display the information graphically (Felsenstein 2004).

Mathematics of Speciation Crow (2009) has suggested that, although until recently mathematical theory had contributed little to the study of speciation, recent mathematical studies by Orr (2005) support it and favor the view that speciation genes correspond to normal genes, selected for their effects within the species. Furthermore, there is evidence that these genes evolve rapidly. Thus, hybrid incompatibility is a by-product of ordinary selection in geographically isolated populations. There is no evidence that random drift plays an important part (Coyne and Orr 2004). The field of mathematical studies of speciation has barely started; it will surely increase. Strangely, Mayr's (1954) theory of "genetic revolutions assumed that genetic drift in a small founder group will occasionally produce a population with a new set of genotypes, although a small founder population seems unlikely to be the source of a new favorable gene combination. This was, in fact, shown to be the case in a mathematical analysis by Barton and Charlesworth (1984).

Reaction to Haldane's "Defense"   Warren Ewens (2008) has been highly critical of Haldane's (1964) defense of beanbag genetics. His main complaint is that it focused too much on minor tactical matters rather than on broad-ranging strategic matters. To quote Ewens (2008),

> One cannot imagine the grand sweep of Newtonian or Einsteinian dynamics without the mathematics involved. Can one imagine the grand sweep of evolutionary theory without the mathematics? To me the answer is clearly "yes": indeed, this is what Darwin produced. But are there key points of the theory that are illuminated and for which mathematics is almost a *sine qua non*, upon which Haldane should have based his defense?(p. 450–451)

I do not agree with Ewens that Fisher would have made a better defense of beanbag genetics than did Haldane. In his "defense," Haldane himself stated that Fisher preferred attack to defense. Furthermore, Ewens (2008) appears to have accepted Mayr's own personal view that theoretical population genetics is based on the "predominantly single-locus mathematical theory" while ignoring genic interactions, epistasis, and other complications. This is far from the truth. Wright included epistasis and other complex situations in developing his shifting balance theory. And so did Haldane in his later papers on the mathematical theory of natural selection. Indeed, in "Metastable Populations," part VIII of his series "A Mathematical Theory of Natural and Artificial Selection," Haldane (1931) investigated the situation where mutant genes are disadvantageous singly but become advantageous in combination. He argued that, for m genes, a population can be represented by a point in m-dimensional space. He also suggested that in many cases related species represent stable types and the process of species formation may be a rupture of the metastable equilibrium, which is more likely to occur when small communities are isolated. In this respect, Haldane came very close to predicting Wright's shifting balance theory, which described evolution as a trial-and-error process in terms of multidimensional adaptive surfaces.

Both Crow (2008) and Ewens (2008) have commented that mathematical methods have become more useful, even indispensable, in recent years with the availability of large volumes of data at the molecular level. Such analyses are retrospective, leading one to inquire about the nature of the evolutionary processes that led to the observed data. Earlier, multilocus evolutionary models had been developed, leading to the concept of linkage disequilibrium, among others.

Especially since the discovery of the Watson-Crick model of DNA, gene action is much better understood, and attention has shifted to regulation from the earlier emphasis on transcription and translation. The importance of what used to be called physiological genetics has increased, and that of beanbag genetics has also increased. This is mainly because, as mentioned earlier, molecular techniques have yielded an abundance of data and the unit of observation, the nucleotide, can be measured accurately and its variability measured with precision.

Population genetics is much more than the "beanbag" genetics that appears to have occupied Mayr's attention inordinately. The genic interaction and cohesion that Mayr was concerned with are now a part of population genetics. Population models now include multiple factors, linkage, dominance, and epistasis. Population genetics can be regarded as an improved beanbag model. It continues the tradition of random drawing of beans from a bag.

Finally, regarding the present situation, Haldane would be pleased by the great success of beanbag genetics. The field is enriched with increased breadth and rigor. Many years ago, in his futuristic book

Daedalus; or, Science and the Future, Haldane (1923) predicted that biology will reign supreme:

> I have tried to show why I believe that the biologist is the most romantic figure on earth at the present day. At first sight he seems to be just a poor little scrubby underpaid man, groping blindly amid the mazes of the ultra-microscopic, engaging in bitter and lifelong quarrels over the nephridia of flatworms, waking perhaps one morning to find that someone whose name he has never heard has demolished by a few crucial experiments the work which he had hoped would render him immortal. There is real tragedy in his life, but he knows that he has a responsibility which he dare not disclaim, and he is urged on, apart from all utilitarian considerations, by something or someone which he feels to be higher than himself.

Cost of Natural Selection

It is Haldane's estimate of the cost of natural selection that epitomizes "beanbag" genetics more than anything else. It also indicates one of the directions in which evolutionary research was moving at that time. "Cost" by definition involves the dynamics of evolution. As Haldane stated, the principal unit process in evolution is the substitution of one gene for another at the same locus, and the substitution of a new gene order, or a duplication, or a deficiency, and so on, is assumed to behave essentially as a unit like a gene in inheritance. Haldane's calculations showed that the number of deaths needed to secure the substitution, by natural selection, of one gene for another at a locus is independent of the intensity of selection. He showed that it is often about thirty times the number of organisms in a generation. If two species differ at 1,000 loci, and the mean rate of gene substitution is one per 300 generations, it will take at least 300,000 generations to generate an interspecific difference.

Haldane's Letters of January 11, 1961 and of March 3, 1961   These letters (see pages 238 and 241) are concerned with the intensity of various "loads." Haldane claimed priority for specifying mutational and segregational loads and for estimating the mutational load in his pioneering paper of 1937. He estimated the mutational load in *Drosophila melanogaster* as about 4 percent, and about 10 percent for humans. Later, both figures were found to be too low. Haldane told Mayr that a "load" even of 50 percent may be "slight" if it occurs early enough in life. Early mortality of zygotes imposes no load on the species' food supply beyond that needed to make half the eggs. On the other hand, if lethals killed the same number during pupation, the load would be much "heavier."

In Mayr's reply of March 16, 1961, we can note the inclusion of certain topics that are close to Mayr's thinking. One is the synergistic action of fitness–reducing factors and the "unrealistic" nature of the assumption that genes have absolute and constant selective values. Mayr argued that various fitness–reducing factors may act in concert, especially through genes causing early mortality, thus eliminating several genes and genotypes without any real threat to the survival of the population. For that reason, Mayr suggested that the cost of natural selection is less severe than Haldane's estimate (1957b).

Mayr emphasized once again that one must go beyond the preliminary assumption that genes have constant and absolute selective values, "an assumption we all know not to be realistic."(Mayr's letter, March 16,1961) Mayr then attacked one of the foundations of Haldane's cost estimate. Mayr wrote:"My feeling is that by operating with 'average selective values' of genes, we introduce quite unrealistic models into our calculations. The mere fact that a lot of the early mortality of zygotes is relatively unimportant for the maintenance of populations is one of the reasons for my doubts."(Mayr's letter, March 16, 1961).

Scaffolding Haldane characterized his mathematical theory as a kind of scaffolding within which a reasonably secure theory which is expressible in words may be built up. Without such a scaffolding, verbal arguments tend to be insecure. Not surprisingly, Haldane provided an analogous example from astronomy, one of the sciences that was often on his mind. When Newton proposed his gravitational theory of planetary movement, some people thought that if the sun attracted the planets they would fall into it. Newton had to show not only that the inverse-square law led to stable elliptic motion, but also that spheres, whose destiny at any point was a function of distance from their center, attracted one another as if they were particles.

Human Genetics   It is curious that Haldane left out the example of human genetics, which is one of the best success stories of beanbag genetics, from his "defense." He did mention that he was leaving out the mathematics of human genetics from his "defense" because one cannot experiment with human species and must extract all the information from available figures. However, it is important to note that the methods that have been so successfully employed in developing the early foundations of human genetics were clearly based on the concepts of beanbag genetics. Haldane did mention his paper on the occurrence of hemolytic disease due to Rh incompatibility and its bearing on the hybrid origin of European populations (Haldane 1942), where his early prediction was vindicated by later evidence. An excellent example of the application of beanbag genetics is the early analysis of ABO blood group genes by Bernstein (1924).

Human population genetics began early. As early as 1902, physician Archibald Garrod showed that some human biochemical disorders such as cystinuria, albinism, and alkaptonuria are determined by autosomal recessive genes, as was evident by the occurrence of parental consanguinity in several patients. Garrod sought the advice of the Cambridge University biologist William Bateson, who coined the term "genetics" in 1906. Bateson confirmed the recessive nature of the biochemical genetic disorders that Garrod was studying at that time. The beanbag nature of Garrod's genetic research is obvious. Indeed, much of early genetics is of the beanbag variety. Fundamental advances in *Drosophila* genetics by Morgan and his colleagues, including the early gene maps by A.H. Sturtevant and others, followed what Mayr later characterized as beanbag genetics.

Formal human genetics began in the 1930s, with the pioneering methods of human pedigree analysis devised by Bernstein (1931), Haldane (1932b), Hogben (1931, 1934), Fisher (1918, 1935), and Penrose (1935), among others. Haldane (1932), for instance, applied Fisher's method of maximum likelihood to estimate the true proportion of recessives in human pedigrees for juvenile amaurotic family idiocy. These early pedigree analyses are clear examples of beanbag genetics. Also see Crow (1993) for Bernstein's studies of the first human marker locus.

Several letters between Haldane and Mayr contain discussions of genetic loads. In his letter of January 12, 1961, Haldane informed Mayr that he was still thinking about the "cost of natural selection" or the "substitutional load" (see p. 238).

Using mathematical modeling was not new to Haldane. Throughout his career, he cited instances of the application of mathematical models in the biological as well as physical sciences. In biochemistry, he worked out the mathematics of enzyme kinetics. It was summarized in his book *Enzymes* (Haldane (1930). He was justifiably proud of his mathematical skills, and enjoyed doing complex calculations by hand. On the other hand, he was equally gifted in scientific popularization, contributing several hundred articles to popular magazines and newspapers. Of the three founders of population genetics, Haldane was the only one who presented popular expositions of the subject in a non-mathematical manner. His great classic, The Causes of Evolution (1932) is a fine example of his style.

2

Foundations of Population Genetics

There are certain aspects of the origins of population genetics that have drawn special attention and comment by the historians of science. First, one wonders about the possible role of personality conflicts in stimulating scientific growth: Do they hinder or help in the development of a branch of science? The antagonism and personal dislike between William Bateson and Karl Pearson delayed the development of genetics, and population genetics in particular, by fifteen years or more. If they had cooperated instead of quarreled, population genetics would have had an earlier start, and it would have developed much further by the time R.A. Fisher, J.B.S. Haldane, and Sewall Wright arrived on the scene. There would be no need for Fisher to write his famous 1918 paper to reconcile the two factions. Other conflicts played different roles. The different approaches and methods followed by the three founders have expanded the wide range of problems, concepts, and methods that have come together to establish that field and shaped its identity. Their conflicts and arguments have enriched the field.

No other field of science was dominated for so long by so few—the three founders, although disagreeing sometimes, developed a cohesive science through their series of papers and books from about 1918 to 1964. Haldane (1964) called the mathematical theory a "kind of scaffolding" within which a reasonably secure theory expressible in words may be built up. He cited examples from his work to show that without such a scaffolding verbal arguments are insecure. It is clear, then, that from Haldane's (1964) point of view, the development of mathematical theory before the collection of any experimental data was not only not surprising but an essential prelude to the development of population genetics.

The following account will show that there is lot more to "beanbag genetics" than was apparent from Ernst Mayr's critique (1959, 1963). It goes far beyond counting Mendelian ratios or a simple genotype/phenotype relationship.

Early Population Genetics

Early developments in genetics could be categorized as "population genetics", much of which could also be characterized as "beanbag genetics" in today's terminology. Indeed, much of classical genetics is based on the assumption that the gene is the basic unit of inheritance for all practical purposes and that there is a direct relationship between a phenotype and its corresponding genotype. The independent nature of each gene and its specific function were generally accepted except where epistasis and linkage were considered. These apparently simple notions have rapidly established classical genetics during the early decades of the twentieth century. One can say that the successful establishment of genetics as a scientific discipline is mainly due to the concepts and methods of "beanbag genetics."

The founder of genetics, Gregor Mendel (see Mendel 1866), himself pioneered "beanbag genetics." It is well known that Mendel's success was due to his clear conception of some new procedures that, according to Bateson (1902), were "absolutely new in his day." One of these was his treatment of individual characters and the underlying factors as discrete noninteracting entities. Among other reasons for

Mendel's success was his clear vision of drawing a simple relationship between a phenotype and its genotype. In other words, these are the same concepts that were followed by Haldane (1924) in formulating his "mathematical theory of natural selection" and to which Mayr (1959, 1963) objected in his critique of "beanbag genetics." It must be added, however, that Haldane took into account the effect of other complications such as epistasis and inbreeding in his later papers in his mathematical series.

Among the early founders of classical genetics, Bateson (1906) in England discovered what was later called "linkage." Bateson introduced concepts and terminology that helped to transform "Mendelism" into the new discipline of "genetics," including the term "genetics" in 1906. In the following years, Bateson introduced other terms: heterozygous and homozygous," allelomorph (which was later shortened to "allele" by Shull), dominant and recessive, and coupling and repulsion (which were later replaced by the biochemical terms "cis" and "trans," as suggested by Haldane in 1941).

Among several others contributing to the foundation of classical genetics were Johannsen (1909). who introduced the term "gene" to replace Mendel's "merkmal" or Bateson's "factor" and made the distinction between a genotype and its phenotype. Nilsson-Ehle (1909) and Emerson and East (1913) introduced the genetic analysis of quantitative characters in maize. Bateson (1909), Morgan et al. (1915), and others who helped found classical genetics assumed a direct relationship between a gene and its corresponding phenotype. Indeed, Mendel himself initiated this line of reasoning. The founders of population genetics, Haldane (1924, 1932a), Fisher (1918, 1930), and Wright (1922, 1931), continued that tradition.

Hardy-Weinberg Law

The "rediscovery" of Mendel's laws in 1900, followed by the independent contributions of Hardy and Weinberg in 1908, led to the expansion and development of population genetics. The Hardy-Weinberg law, as it came to be called later, is the generalization that the stable frequency of genotypes is p^2AA : $2pq$Aa : q^2aa. Hardy (1908) considered his contribution so trivial that he did not even mention it in his autobiographical account, "A Mathematician's Apology."[1] Earlier, he wrote: "I should have expected the very simple point which I wish to make to have been familiar to biologists" (Sturtevant 1965, p. 108). But it was not, because it had been seriously suggested that dominant genes would automatically increase in frequency in mixed populations!

The Hardy-Weinberg law is strictly valid only if several conditions are fulfilled:

(1) The population must be large enough that sampling errors can be ignored.

(2) There must be no mutation.

(3) There must be no selective mating.

(4) There must be no selection.

With the second and fourth requirements, we should note the possibility that balanced mutation and selection rates may exist, resulting in no net changes in the frequencies of A and a.

The Mendelian–Biometrician Controversy

William Bateson (1861–1926) attained fame as the outspoken Mendelian antagonist of Walter Raphael Weldon (1860–1906), his former teacher, and Karl Pearson, who led the biometric school of

[1] It has been alleged that Hardy, as a distinguished mathematician, was so embarrassed at the triviality of his contribution to genetics that he decided to publish the paper in *Science*, a distant journal across the Atlantic, where it would not be seen by his mathematical colleagues in England!

thinking. This concerned the debate over saltationism versus gradualism. (Darwin had been a gradualist, but Bateson was a saltationist.)

Later, R.A. Fisher and J.B.S. Haldane showed that discrete mutations were compatible with gradual evolution. The biometricians
rejected Mendelian genetics on the basis that discrete units of heredity, such as genes, could not explain the continuous range of variation
seen in real populations. Weldon's work with crabs and snails provided
evidence that selection pressure from the environment could shift
the range of variation in wild populations, but the Mendelians maintained that the variations measured by biometricians were too insig-
nificant to account for the evolution of new species.

The Mendelian and biometrician models were eventually reconciled
with the development of population genetics. A key step was the work
of Fisher, a British statistician. In 1918, Fisher's landmark paper "The
Correlation between Relatives on the Supposition of Mendelian
Inheritance," written in 1916, laid the foundation for what came to be
known as biometrical genetics and introduced the very important
methodology of the analysis of variance, which was a considerable
advance over the correlation methods used previously. The paper showed
very convincingly that the inheritance of traits measurable by real values,
the values of continuous variables, is consistent with Mendelian principles. It helped to resolve the long-running dispute between the
Mendelians, led by Bateson, and the biometricians, led by Pearson.

**The Founders of Population Genetics: Haldane, Fisher, and
Wright**   The pioneering mathematical studies of evolution by Haldane (1924, 1932a), Fisher (1930), and Wright (1922, 1931) laid down
the foundation for a meaningful interpretation of evolutionary phenomena in the context of Mendelian genetics. However, the first book
under the title "population genetics" was not published until the 1940s,[2]
although the term itself was used earlier by Theodosius Dobzhansky
(1937).

---

[2] A book by the Chinese geneticist Ching Chung Li with the title *Population Genetics*
was published in China in 1946.

The importance of the mathematical contributions of Fisher
(1931), Haldane (1932a), and Wright (1931) was almost universally
accepted by biologists for more than three decades. Later, two promi-
nent biologists, C.H. Waddington (1957) and Ernst Mayr (1959),
began questioning the value of the contributions of these great pioneers. It is of interest to examine their criticisms and Haldane's (1964)
"defense" (see chapter 1). It also appears to be necessary to explain
the origins and meaning of "beanbag genetics," because certain misconceptions still abound.

In his "defense of beanbag genetics," Haldane (1964, p. 357) wrote:

> Sewall Wright has been the main mathematical worker in this field,
> and I do not think Mayr has followed his arguments. Here Wright is
> perhaps to blame. So far as I know, he has never given an exposition of
> his views which did not require some mathematical knowledge to
> follow. His defense could be that any such exposition would be
> misleading. I have given examples above to illustrate this possibility
> (Haldane 1964, p. 357).

Haldane's "defense" was clearly too little and too late. During the
preceding years, when Waddington and Mayr were criticizing the
contributions of theoretical population genetics, there was no response
from Fisher, Haldane, or Wright. None of them took the trouble to
respond to the critics. Their names and contributions were so well
established in the genetic literature that it might have seemed to them
almost unnecessary to respond to their critics.

As far as popular exposition is concerned, Haldane was the only
one who presented nonmathematical summaries of his theories, for
instance, in *The Causes of Evolution* (1932a), but it was written long
before Waddington (1957) and Mayr (1959) questioned the value of
theoretical population genetics. Both Fisher's and Wright's publications and lectures were addressed to professional geneticists or, more
accurately, a very small fraction of that community, because most
geneticists were not mathematically trained to understand these
papers. Haldane's popular expositions (1932a, 1955, 1959), although

extremely lucid and largely nonmathematical, were not a "defense" of theoretical population genetics. But they have educated many scientists and layman alike by providing a basic understanding of the mathematical theory of natural selection and the major issues involved therein. Fisher and Wright provided no popular expositions.

Delayed Response   There was clearly a delay of several years from the time when Waddington and Mayr began questioning the value of the contributions of the mathematical school and Haldane's (1964) "defense." This appears to be inexplicable at first glance. However, Haldane (1964) offered an explanation of sorts, at least to account for the silence of Fisher and Wright: "Fisher is dead, but when alive preferred attack to defense. Wright is one of the gentlest men I have ever met, and if he defends himself, will not counterattack. This leaves me to hold the fort, and that by writing rather than speech." (p. 344) But Haldane offered no explanation for his own tardiness in responding to the criticisms of Waddington and Mayr.

There is another reason that could account for their nonresponse. From the nature of Haldane's (1964) tongue-in-cheek "defense," it is clear that he regarded Mayr's knowledge of mathematics as well as mathematical genetics as quite inadequate to indulge in a serious debate of any kind. It is very likely that Fisher, and Wright also, never took the criticisms seriously, certainly not worthy of a serious response! They had nothing to worry about: The contributions of all three were widely recognized and established. As for Haldane, he never missed an opportunity for a good fight or a controversy!

Different Approaches   The mathematics of all three pioneers, Haldane, Fisher, and Wight, agreed essentially despite some differences in their approach. Haldane expressed his results in terms of the ratio, u, of the frequency of the mutant gene (A^1) to that of its type allele (A), (distribution $1A : uA^1$). Fisher (1922a) preferred a function of the gene frequency distribution $(1 - p)A : pA^1$, namely, $\theta = \cos^{-1}(1-2p)$, which has the property that its sampling variance is constant $1/(2N)$, but he shifted later (1930) to the gene frequency itself, which was used systematically by Wright (1929, 1931). There were, however, great differences in the application to evolution.

J.B.S. Haldane (1892–1964)

I have provided information about Haldane's biographical details elsewhere in this book, as have other sources (I recommend; Crow 2004; Dronamraju 1968, 1985, 1990, 1995). Haldane assigned selective values, usually constant, sometimes variable, to each gene or, in some cases, each genotype involving two or more interacting loci. He deduced deterministically the number of generations required to bring about a specified change in the gene frequency ratio under various alternative assumptions regarding the genetics of the character.

In the first of his series of mathematical papers on evolution, Haldane (1924) wrote that a satisfactory theory of natural selection must be quantitative. He wrote: "In order to establish the view that natural selection is capable of accounting for the known facts of evolution, we must show not only that it can cause a species to change but that it can cause it to change at a rate which will account for present and past transmutation." (p. 19) In order to determine how the frequency of a gene in a population changes under natural selection, Haldane (1924) made several simplifying assumptions that are listed on pages 14–15.

Haldane (1924, p. 57) specified that the following information must be obtained first:

(a) The mode of inheritance of the character considered

(b) The system of breeding in the group of organisms studied,

(c) The intensity of selection,

(d) Its incidence (e.g., both sexes or only one), and

(e) The rate at which the proportion of organisms showing the character increases or diminishes.

It should then be possible to obtain an equation connecting (c) and (e) . Haldane stated at the outset that he would be dealing only with the simplest possible cases, involving a single completely dominant Mendelian factor or its absence. In one instance, considering the effects of slow selection, Haldane (1924) showed that for an autosomal factor (he was still using the term "factor", which was introduced by his mentor Bateson and later replaced by "gene"), the number of generations required to change the frequency, under slow selection ($k = 0.001$), are as follows:

0.001% to 1.0%	6,921
1.0% to 50.0%	4,592
50.0% to 99.0%	4,592
99.0% to 99.999%	6,921

When external conditions change drastically, many genes that have been less favorable than their type alleles become more favorable, and the more deterministic process of Haldane (and Fisher) dominates the situation until there is approximate adjustment to the new conditions. The concrete cases to which Haldane applied his theory were ones that involved such change. One of these cases (Haldane 1942) was based on data assembled by the ecologist Charles Elton on the steady decline in the proportion of silver fox pelts among fox pelts marketed in various parts of Canada in the century preceding 1933. Silver is due to a simple recessive gene. Haldane calculated that this gene must have been at an average selective disadvantage of about 3% per year or 6% per generation if generation length is taken as two years, compared with it allele in red.

Another case involved Haldane's (1924, 1956) analysis of the nearly complete replacement of light-colored moths, *Biston betularia*, by a semidominant dark mutant form (which in the end became completely dominant, presumably by direct selection of modifiers) in industrial districts of England in the course of half a century. Haldane found that the selective value of the original light form must have averaged only two-thirds that of the dark, and much less at times and places in which selection was most severe, an estimate that seemed to some to be excessive when first made in 1924. Haldane's estimate was in accord with the later observations of Kettlewell (1956) on the relative amounts of destruction by birds of the two color varieties on soot-covered, compared with clean, tree trunks.

If change of environment keeps pace with major changes in gene frequencies, evolution according to Haldane's theory may continue indefinitely, and a case can be made for this as a major evolutionary process. This is a cyclical process, each new adaptation is followed by the undoing of an old one. The process is deterministic only during each period of adjustment, since the changes in conditions on which continuance of the evolution depends introduce an indeterminate aspect. The situation is unfavorable for either type of theory if external conditions are stable over very long periods of time and population structure is unfavorable for the stochastic process. Such evolution as there is would be of Haldane's type, limited by the exceedingly rare occurrence of novel (nonrecurrent) favorable mutations. If population structure is favorable and external conditions are changing at a tempo compatible with evolutionary adjustment, a compound process should occur that probably leads to more rapid evolution than either by itself. In a later paper, Haldane(1949a) recognized the importance of fine-grained subdivision of the species.

Haldane's contributions to evolutionary theory include comprehensive discussions, both verbal and mathematical, of many aspects, including the evolution of dominance (Haldane 1930, 1939), a subject introduced by Fisher in 1928, selection in relation to the time of action of the gene in the life cycle or reproductive cycle (Haldane 1932c), the evolutionary consequences of recurrent mutation (Haldane 1933), and residual heterozygosis in pure lines (Haldane 1936). To these must be added Haldane's (1937) study of the reduction in fitness when equilibrium is attained between recurrent mutation and selection compared with fitness in the complete absence of the mutant gene (also see Haldane 1927, a,b). In this instance, Haldane arrived at a simple principle (sometimes known as Haldane's principle) that the fitness at equilibrium, $1 - v$, for a completely recessive mutation or $1 - 2v$ for a more or less dominant one, depends merely on the rate, v, of recurrence of the mutation and not at all on the severity of selection (although, of course, the latter has a great deal to do with the number of generations required to approach equilibrium after a change in mutation rate).

This principle has been interpreted as implying that practically all mutations (from barely deleterious to lethal) are essentially equivalent in the load they impose on the population, an interpretation that was useful in estimating the genetic damage to populations resulting from high-energy radiation from nuclear explosions and other sources. Haldane himself, however, stressed mostly lethal or nearly lethal mutations because of the rapidity with which the genetic damage is manifested. He proposed methods to determine the rate of induction of such mutations in mice from linked marker genes (Haldane 1956, 1960). He suggested (1955) that the effects probably would be much more serious than had been estimated on the basis of experiments with *Drosophila* because of the likelihood that nearly all spontaneous mutations in long-lived forms, such as humans, are due wholly to high-energy radiation (instead of a small fraction of 1.0% in *Drosophila*), as a result of evolutionary protection from low-energy agents that otherwise would prevent all reproduction. Haldane suggested that the amount of radiation required to double the spontaneous mutation rate in the human species might well be as low as 3 roentgens instead of 50–80 roentgens, as deduced from *Drosophila* experiments.

Among other evolutionary topics analyzed by Haldane were the theory of clines (1948) and various mechanisms that may be responsible for the presence of two or more alleles in a population (Haldane 1942, 1954, 1955; Haldane and Jayakar 1963; summarized in Dronamraju 1968) by various contributors. Several papers of Haldane in evolutionary biology were reprinted in *Selected Genetic Papers of J.B.S. Haldane* (Dronamraju 1990). Other contributions of Haldane include the quantitative measurement of the rate of evolution (1949c) and the measurement of the severity of natural selection (1954, 1959).

Types of Selection Haldane (1959) provided an excellent summary of various types of selection on the basis of the nature of consequences resulting in each situation. Selection that alters the mean of any character may be called *linear*. If such selection reduces the variance of a character, thereby weeding out extreme phenotypes, it can be called *centripetal*. If it increases the variance, it is called *centrifugal*. Karn and Penrose (1951) measured the intensity of selection for human birth weight. Such selection is largely centripetal because it involves a reduction of phenotypic variance (by about 10 percent). On the other hand, centrifugal selection occurs rarely, for instance in situations when a new polymorphism is being established. Haldane (1957b) cited Kettlewell's (1956) results on selection for melanism as an example of both linear as well as centrifugal.

Haldane distinguished between *effective* and *ineffective* selection: The former changes the gene frequencies, whereas the latter does not. For instance, selection favoring heterozygotes may be effective for a while but will lead to a stable equilibrium when it is ineffective. On the other hand, selection based on negative heterosis is always effective, because it results in the elimination of one allele or the other. Selection against mutants is ineffective when it is only balancing mutation.

Mather (1953) and Waddington (1953) have contributed to the terminology of evolution. If the mean of a character changes in a certain expected direction (and is not directly due to environmental change), then this may be called *directional* evolution. If the variance is reduced, by eliminating those genotypes whose means differ widely from the population mean, then it may be called *normalizing* evolution. If its reduction involves elimination of genotypes, which vary greatly in diverse environments, then we may call it *stabilizing* selection. On the other hand, if the variance is increased, it is called *disruptive* evolution. Haldane (1957b) expressed some reservation concerning the term "disruptive" in this context because he did not consider a process disruptive when it involves the establishment of a stable polymorphism.

It might be noted further that at equilibrium evolution is neither stabilizing nor normalizing. Were it not for the fact that selection against most mutants is constantly occurring, the mutations would result in disruptive evolution. Such selection is always centripetal and also may be linear.

Haldane (1959) emphasized that a separate vocabulary is needed for the different types of phenotypic and genotypic selection. He cited the example of selection at the human *Rh* locus due to neonatal jaundice caused by the D antigen, which is phenotypically disruptive but genotypically directional. In order to measure "phenotypic selection," Lush (1951) suggested the term "selective differential" for the difference between the mean value of a character in the parents of the next generation and the mean in the population from which they are drawn. Another term, "selective advance," refers to the difference between such means in successive generations. Such methods are very useful in studying artificial selection.

Haldane (1954) suggested a different measure.

If s_0 is the fraction of the optimal phenotype surviving, and S is the (smaller) fraction surviving of the total population, Haldane defined the intensity of selection as

$$I = \log_e s_0 - \log_e S$$

This measure of Haldane's is quite different from that of Lush (1951). Unlike Lush's measure, it is applicable to the case of natural selection, which may have no effect on the mean, but may reduce variance quite considerably.

Cost of Natural Selection   One of his most brilliant and influential papers in population genetics was published during the last years of his life. In a discussion of what he called the "cost of natural selection," Haldane (1957b) estimated the cost of replacing an allele in a species by a new one, that is to say, the cost to the effectiveness of selection at other loci during the course of evolution. It was

Haldane's analysis of the cost of natural selection that formed the main justification for Motoo Kimura's (1968) "neutral theory of evolution." Haldane investigated mathematically the cost paid by a species per generation during the course of adapting, through natural selection, to changed circumstances such as a change in the environment, an alteration in climate, a new predator, a new source of food, or migration to a new habitat. The cost to the species is expressed in differential mortality and lowered fertility. Referring to a particular locus, Haldane showed that the total number of selective deaths (or the equivalent in lowered fertility) depends mainly on the initial frequency of the gene that subsequently is favored by natural selection. Haldane showed that the cost incurred by the species, during the process of gene substitution, could be as high as thirty times the population number in a single generation. The number 300 generations is a conservative estimate for a slowly evolving species not at the brink of extinction by Haldane's calculation. For a difference of at least 1,000 genes, 300,000 generations might be needed—maybe more. Haldane (1957b) also stated that the rate of evolution is set by the number of loci in a genome and the number of stages through which they can mutate. However, Mayr (1963, p. 400) disagreed with Haldane because elephants and other slow-breeding mammals seem to have evolved much faster than *Drosophila*, which has numerous annual generations, and suggested that other factors play a more important role than the number of loci.

The term "cost" also reminds us of "bookkeeping," which Gould (2002, p. 634–635) emphasized while referring to Crow's (1994) analysis of the foundations of population genetics. Crow (1994) wrote:

> The pioneers ... and their intellectual heirs have been concerned, not with selection as an end in itself, but with selection as a way of changing gene frequencies. Selection acts in many ways: it can be stabilizing; it can be diversifying; it can be directional; it can be between organelles; it can be between individuals; it can be between groups.... But the bottom line has always been how much selection changes allele frequencies and through these, how much it changes phenotypes. This suggests that

we should judge the effectiveness of selection at different levels by its effects on gene frequencies. (p. 616)

Gould's interpretation is that these alterations in allelic frequencies should be read as a "bottom line" in judgments about selection's effect. The notion of "bottom line" also brings forth genes as units of bookkeeping.

Haldane's Dilemma The first use of the term "Haldane's dilemma" was by paleontologist Leigh Van Valen in his 1963 paper "Haldane's Dilemma, Evolutionary Rates and Heterosis." Kimura (1960, 1968) has referred to Haldane's "cost" as the substitutional (or evolutional) load. It is a dilemma for the population: For most organisms, rapid turnover in a few genes precludes rapid turnover in the others. A corollary of this is that, if an environmental change occurs that necessitates the rather rapid replacement of several genes if a population is to survive, the population becomes extinct.

> Since a high number of deaths are required to fix one gene rapidly, and dead organisms do not reproduce, fixation of more than one gene simultaneously would conflict. Note that Haldane's model assumes independence of genes at different loci; if the selection intensity is 0.1 for each gene moving towards fixation, and there are N such genes, then the reproductive capacity of the species will be lowered to 0.9^N times the original capacity. Therefore, if it is necessary for the population to fix more than one gene, it may not have reproductive capacity to counter the deaths. (Van Valen 1963, p.523)

Haldane (1957b) concluded his paper by acknowledging that his conclusions needed to be revised: "To conclude, I am quite aware that my conclusions will probably need drastic revision. But I am convinced that quantitative arguments of the kind here put forward should play a part in all future discussions of evolution."

Haldane (1957b) and Kimura (1960) agreed in two important aspects: (1) The replacement of one gene by another is a slow process because of the high number of genetic deaths resulting from the process, and (2) the number of loci at which genes can be substituted simultaneously is low. Kimura (1960) estimated that not more than about a dozen loci can be involved at any one time, or else the survival of the species is seriously threatened.

Mayr's Critique   Mayr (1963) offered the following criticisms regarding Haldane's "cost" estimate.

(a) Mayr suggested that the contributions to genetic death made by each gene substitution may not be completely independent of each other, as was implied in Haldane's original paper (1957b). It is quite likely that their actions are partially synergistic in a genotype.

(b) Selection coefficients are usually treated as constants. However, they do change their values from one generation to next, depending on various other factors, although it is quite possible that the overall fitness of a population over long periods remains constant and that the addition of superior genes or the elimination of deleterious genes alters the selective values of the remaining genes. Mayr suggested that this effect will mitigate the impact of gene substitution and permit an increase in the rate of substitution.

(c) Haldane mentioned the importance of density-dependent factors, but it is difficult to estimate their effect. Any mortality occurring will have less effect on the fitness of the population as a whole if it occurs earlier in the life cycle. Mayr pointed out that if gene substitution results in a higher rate of mortality, the "ecological load" will be lighter on the remainder of the population. He believed that the survival of the population is guaranteed by this feedback mechanism, despite the increased substitution load.

(d) Haldane showed that the cost to a population of making a gene substitution depends only on the natural logarithm of its initial frequency, as long as selection is not too intense. In a general hypothetical case, one assumes that if the population is very large, selection intensities are low, and the initial frequency is determined by mutations, this would lead to extremely slow evolutionary rates. However, it is easy to show that none of these conditions may exist in natural populations.

None of these possibilities invalidates Haldane's and Kimura's basic approach. However, the numerical values suggested by them for rates of gene substitution (and evolution) will require careful revision.

Natural Selection   Haldane's (1959) essay "Natural Selection" was a contribution to the book *Darwin's Biological Work: Some Aspects Reconsidered*. It was written at the time of his move to India in 1957 and was published two years later. It was also a reflection of how Haldane's methods of beanbag genetics have led to an understanding of natural selection, utilizing data collected in nature.

Haldane's (1959) essay was a largely nonmathematical summation of his view of natural selection, a brilliant exposition of the relationship between Darwinian evolution and Mendelian genetics. He pointed out that the "most important advance in knowledge bearing on Darwinism in the last century has been the elucidation of the fundamental laws of genetics." Remarkably, Haldane proceeded to explain how he would educate Darwin if he could be resurrected for a week. He wrote:

> I should recommend him to read Mayr's <u>Systematics and the Origin of Species</u>, and Simpson's <u>The Major Features of Evolution</u>. He would find most of the facts new and thrilling. But he would understand most of the arguments. I would not recommend him to read Dobzhansky's *Genetics and the Origin of Species* (1937). He could not profitably do so without a grounding in elementary genetics. (p. 104)[3]

Haldane began by presenting a comprehensive "sketch of genetics," which he thought was necessary before discussing various aspects of

---

[3] In a review of that book, edited by P.R. Bell, Dobzhansky (1960) remarked that he would recommend Darwin's ghost to read Haldane's chapter on natural selection.

evolution by natural selection. It was an excellent example of "bean-bag genetics."

For the present purpose, Haldane's (1959) discussion of selection is of interest. After considering various types of selection such as directed selection, centripetal selection, and centrifugal selection, Haldane estimated the intensity of selection where the targets of selection considered are both the phenotype and the genotype. The simplest case considered by Haldane is a hypothetical population where generations do not overlap, as in certain annual plants and animals. Even simpler examples are organisms with only one parent, as in apogamous plants like the dandelion (*Taraxacum officinale*) or self-fertilized hermaphrodites like the pea (*Pisum sativum*).

Biston betularia   Haldane then considered the well-known case of directed natural selection involving the peppered moth, *Biston betularia*, which was discussed in his "defense" (1964). Many species of Lepidoptera have become dark in the industrial areas of England that are polluted by industrial smoke. In only about sixty years after the first melanics appeared, they had replaced the original type completely. The almost black variety, carbonaria, is found over a wide area of England where the lichens on tree trunks have been killed by smoke and the bark more or less blackened. It is due to a single dominant gene C. Another variety, insularia, a less dark variety, was found in partly polluted areas. Haldane (1924) showed that this rapid change could not be due to mutation, which would have to be more than 10 percent.

Haldane observed that in some or even most of these cases, centripetal as well as directional selection is occurring. He speculated that it might take several more centuries before CC became as fit as Cc by the selection of modifiers. However, in the meantime, the evolutionary trend could be reversed if the smoke pollution comes to an end.

In his "defense" of beanbag genetics, Haldane (1964) drew attention to a curious error on page 191 of Mayr's (1963) book *Animal Species and Evolution*. Mayr noted that Haldane's (1932a) "classical" calculations were deliberately based on very small selective intensities and implied that Haldane reached the same conclusion for industrial melanism only in 1957. Mayr (1963, p. 191) wrote:

> The classical calculations of the power of natural selection (Fisher 1930, Haldane 1932a) were deliberately based on very slight differences in the selecting value of competing genes, in order to demonstrate that evolutionary changes would occur even when one gene was superior to its allele by as little as 0.1 percent. Much evidence is now accumulating to indicate that the differences in selective values among genotypes occurring in natural populations may be as high as 30 or 50 percent.

In fact, it was not until 1957 that biologists took Haldane's 1924 calculations seriously, when he reported that it conferred a selective advantage of about 50 percent on its carriers. Haldane (1932a) did not stress such intense selection in *The Causes of Evolution* because it was so unusual that he thought it was unimportant for evolution. Haldane (1964, p. 348) wrote: "If biologists had had a little more respect for algebra and arithmetic, they would have accepted the existence of such intense selection thirty years before they actually did so."

Haldane concluded that most supposed examples of natural selection are very hard to detect. A well-known case, cited by Haldane, was Darwin's interest in the occurrence of wingless insects on oceanic islands. Darwin thought this was mainly because the winged forms were being blown out into the ocean. Haldane commented that no one had introduced a winged species onto such an island to see if it passes though a period when some members are wingless. It is further well known that some insects have developed resistance to such chemical compounds as DDT. Haldane concluded that such instances certainly represent Darwinian selection, not a Lamarckian effect.

Darwin   Darwin was unique among his contemporaries in giving the rate of evolution approximately correctly. In the first edition of

The Origin, he estimated the age of the Upper Jurassic as 300 million years on the basis of a simple calculation about the rate of erosion. A modern estimate based on radioactive minerals is about 130 million years. That was the time when some of his contemporaries were estimating the age of the earth at under 6,000 years, a value based on ancient Hebrew documents. However, Sanskrit documents from ancient India estimated it to be about 4,000 million years, which is closer to the modern scientific estimates.

Simpson (1944) gave two types of estimate of the rate of evolution. The average duration of a genus of carnivores is about 8 million years, and that of a species is less than 1 million. There are vastly differing rates of evolution. The mollusks were much slower, and a few molluscan genera are more than 300 million years old. Another kind of estimate is based on the rate of evolution of quantitative characters such as tooth size in the ancestors of present-day horses. Haldane found that when linear measurements are considered, the average increase in mean measurements was only abut 3.5 percent per million years.

Haldane (1949c) suggested the term "Darwin" for the unit of rate of increase or decrease by a factor of e (2.18) per million years, or by approximately a factor of 1.001 per 1,000 years. The average rate for horse teeth was about 30 to 40 millidarwins.

As Haldane pointed out, if no other agency is opposing selection, the rate of evolution is proportional to the intensity of selection when it is fairly small, about less than a 10 percent advantage to the favored phenotype. Centripetal selection generally weeds out extreme phenotypes. However, weak selection will seldom overcome centripetal selection.

Haldane believed that Darwin was justified in arguing from artificial to natural selection. But it was a weak argument. Darwin was not aware that there were stabilizing agencies that could prevent the occurrence of evolution, leading to stagnation or extinction.

A Detailed Analysis of *The Causes of Evolution*   Haldane (1932a) in *The Causes of Evolution* integrated both biochemical genetics and cytogenetics. Carson (1980) appeared to have grasped the wide coverage found in *The Causes*, which included a synthesis of several growing points of biology at that time: "Haldane neatly conjoins Darwin and Mendel, Fisher and Wright, Newton and Kihara. In the evolutionary context, Haldane deals for the first time with inversions and translocations, polyploidy and hybridization. The paleontological record is woven into the argument."

With reference to the relationships between various animal forms, Haldane (1932a) speculated that a biochemist who finds the same quite complex molecules in all plants and animals can hardly doubt their common origin. He wrote further: "There may be some reason in the chemical nature of things why all living creatures must contain glucose." (p. 79) One finds Haldane's biochemical speculations in several other publications.

As early as 1920, Haldane was discussing the gene in physico-chemical terms, "The chemist may regard them as large nucleoprotein molecules, but the biologist will perhaps remind him that they exhibit one of the most fundamental characteristics of a living organism: they reproduce themselves without any perceptible change in various different environments." (p. 448)

Why Was The Causes of Evolution Less Quoted? *The Causes* was less often quoted than the works of Fisher (1930) and Wright (1931) in the scientific literature. There are at least three reasons that can account for this:

(a) This may be in part due to the fact that some writers (e.g., Provine in Sarkar 1992) interpreted the development of mathematical population genetics primarily as an argument between Fisher and Wright, with Haldane playing a quite role in the background.

(b) Mayr (1992) suggested that the paucity in references to *The Causes of Evolution* may have been due to it being considered a somewhat popular work and "was less often referred to in the United States in the technical evolutionary literature than the works of Fisher and Wright, and the volume was missing in many libraries" (Mayr 1992, p. 175).

(c) At least one reviewer of my book of Haldane's collected papers (Dronamraju 1990) remarked that many fellow scientists in his generation were not referred to study Haldane's books and papers, perhaps because of Haldane's communist and Marxist activities.

The Causes was devoted, in part, to refute several misconceptions and false beliefs in the public mind about the theory of evolution. Haldane was also concerned with correcting Darwin's errors, such as his belief in the theory of blending inheritance.

He pointed out that acceptance of blending inheritance would require a huge reduction of variation in every generation and would require the production of an equally colossal new genetic variation in every generation. Darwin mistakenly thought that this could be accomplished by the environment. Haldane (1932a) emphasized that Darwin had no knowledge of the main causes of variation within a species, such as mutation, recombination, and chromosomal aberrations.

In the early 1930s, when Haldane was writing *The Causes*, he made a serious effort to refute all claims of an inheritance of acquired characters, because Lamarckian inheritance was still rather popular in some circles. He explained that some experimental results that were supposed to provide evidence in favor of the inheritance of acquired characters were, in fact, due to unconscious selection. And, it is only a matter of time until all such claims are refuted. Such a refutation of Lamarckian inheritance was still highly necessary at that time (Mayr 1992). Indeed, Haldane performed that task so well that very little space was devoted to this topic only seven years later by Dobzhansky when he wrote *Genetics and the Origin of Species* (1937).

Haldane (1932a) also dismissed two other theories competing with Darwinism in the presynthesis period: orthogenesis and salta-tionism. Because of the great popularity of orthogenesis at that time, Haldane (1932a) took great pains to emphasize that "the study of evolution does not point to any general tendency of a species to progress" (p. 153). On the contrary, he pointed out that "Most lines of descent end in extinction, and commonly the end is reached by a number of different lines evolving in parallel" (1932a, p. 85).

Referring to regress and degeneration, Haldane wrote: "The change from a monkey to man might well seem a change for the worse to a monkey. But it might also seem so to an angel (p. 83) and "monkey is quite a satisfactory animal. Man of to-day ... is a worse animal than the monkey"(p. 83). He was referring to the difficulties in human structure caused by upright posture and bipedalism.

Saltationism and Typological Thinking It is of interest to examine Mayr's evaluation of Haldane's *The Causes*. Mayr (1992) wrote that *The Causes* was written at a time when most authors considered mutation to be a deleterious process. For instance, Haldane (1932a) wrote: "Most mutations lead to a loss of complexity ... or reduction in the size of some organ.... This is probably the reason for the at first sight paradoxical fact that, as we shall see later, most evolutionary change has been degenerative." (p. 83) And with regard to the final conclusion on the importance of mutation, Haldane stated: "And if the relative importance of selection and mutation is obvious, it has certainly not always been recognized as such." Mayr (1992) commented, "An obvious reference to the Mendelians."

Mayr noted that the saltationism of the Mendelians clearly reflected their typological thinking and that Haldane (1932a) had gotten rid of this to a large extent. Yet, according to Mayr (1992), some of Haldane's (1932a) statements on mutations were still tainted with typological thinking, and even more so were some of his statements on species. However, Mayr (1992) admitted that Haldane argued against the belief of the Mendelians in the importance of mutation pressure, as a factor that could override natural selection. Haldane (1932a) argued that selection can almost always overcome any effects of mutation and that "we cannot regard mutation as a cause likely by itself to cause large changes in a species." (p. 60)

Based on estimates of mutation rates in the fruit fly *Drosophila* and in the plant *Primula sinensis*, Haldane (1932a) estimated that the frequency of mutation for hemophilia by assuming that "the rates of production by mutation and elimination by natural selection must about balance, and the probability of mutation of the normal gene works out at about 10^{-5} per life-cycle" (p. 32). Regarding the importance of mutation in evolution, Haldane wrote: "It enables us to escape from the impasse of the pure line. Selection within a pure line will only be ineffective until a mutation arises." (p. 32)

Haldane (1932a) was concerned that the so-called "mechanism of heredity" might be interpreted in some quarters as the dawn of "atomism" in biology and that Mendel reduced genetics to biophysics. Haldane (1932a) wrote:

> It is at present irrelevant to genetics whether life is or is not ultimately explicable in terms of physics and chemistry.... Genetics can give us an explanation of why two fairly similar organisms, say a black and a white cat, are different. It can give us much less information as to why they are alike. In the same way a complete theory of evolution might tell us how the different various different species had originated from common ancestors. But it would give us little direct information concerning the nature of life. (pp. 33–34)

Species   Referring to Haldane's *Causes*, Mayr (1992) wrote: "For him a species was largely an aggregate of genes. And, like Darwin, he was much more concerned with showing that species are related to each other and that one can derive one species from another, than to show how this occurs." (p. 178) In chapter 3 of *The Causes*, Haldane demonstrated that the same gene may occur in different related species, just as Nikolai Vavilov had done previously for cultivated plants. Certain characters of two parental species segregate in the F2 generation in all cases where it is possible to cross species.

Mayr (1992) continued his criticism: "Yet, in this account he virtually ignores the fact that a large part of the phenotype of species hybrids shows 'blending' in the F2; i.e. it does not show Mendelian inheritance." (p. 178) Mayr argued that this phenomenon was used at that time by anti-Mendelians to support the existence of some sort of a "species substance" that blended rather than segregated. He added,

"This belief was never specifically refuted by Haldane, but he mentioned that 'quantitative' differences could be explained in terms of multiple genes affecting the same character." Mayr (1992) appears to have misunderstood Haldane's position on this point.

Haldane attempted to explain how Darwin may have reached his conclusions regarding blending inheritance, but there was no suggestion in any sense that Haldane was accepting blending inheritance. For instance, Haldane (1932a) wrote: "Darwin observed blending, i.e. diminution of variation, because the mating system of his domestic animals and plants was suddenly changed, for example, when two races of pigeons, which had been bred separately for many years, were mated together" (p. 8). Indeed, Haldane (1932a) emphasized the contributions of both Mendel and Bateson to the understanding of segregation:

> "The tendency to blending is exactly balanced by the opposite process of segregation, by which the offspring of a given union vary among themselves in respect of heritable characters. But whereas the phenomenon of heredity had been known in a general way for ages, that of segregation was first seriously studied by Mendel, in the nineteenth century, and it was above all Bateson who stressed its significance as a biological fact as important as heredity. (p. 8)

Haldane emphasized that the kind of genetic differences that exist between related species also exist within species.

Such early Mendelians as Bateson, Hugo DeVries, and Johannsen accepted that evolution occurred through major mutations. Haldane (1932a) quoted the botanist John Willis (1922), who applied the saltationist approach to evolution in his book *Age and Area*. Although Haldane (1932a) did not believe that saltations played any significant role in the evolutionary process, he was impressed by Willis's estimate that, in flowering plants, the number of new successful species appearing is about two per century. Haldane (1932a) noted that the role of natural selection is to decide which of the new species formed by mutation will survive.

At the time when Haldane wrote *The Causes of Evolution* (1932a), there was a great need to clear the air by getting rid of several false beliefs about evolution that still lingered from the past: Inheritance of acquired characters (Lamarckism), orthogenesis, saltationism, excessive importance of mutation pressure in evolution, and creationism, to name a few. He performed that task admirably. With respect to the role of mutation pressure, Haldane (1932a, p. 109) wrote that it must be a slow cause of evolution but that it cannot be neglected, especially when organisms are in a fairly constant environment over long periods. It will favor polyploids, especially allopolyploids, which possess several pairs of sets of genes, so that one gene may be altered without disadvantage, provided its functions can be performed by a gene in one of the other sets of chromosomes.

In his review of the reprint of Haldane's *The Causes of Evolution*, which was reprinted in 1990, Mayr (1992) praised Haldane for his open mind; he wrote: "I have always admired Haldane for the frankness with which he acknowledged his ignorance about certain problems.... Haldane was the most open-minded of the Fisher-Haldane-Wright trio. He perceived aspects of evolution that are often ignored." (p. 179)

It should be noted further that *The Causes* was written at a time when the role of natural selection in evolution was either poorly understood or ignored. With regard to selection, Haldane wrote that we know very little about what is being selected and that much of the opposition to natural selection is due to "a failure to appreciate the extraordinary subtlety of the principle of natural selection." Haldane emphasized that one should look beyond external morphology when studying the process of evolution. More important are physiological adaptations that permit organisms to perform their daily functions and to adjust to life in different climatic zones or in specialized habitats. Certain aspects of developmental physiology, such as those that affect the rate of embryonic growth, might also be very important.

Is Mathematical Analysis Necessary?   In the concluding paragraph of *The Causes of Evolution*, Haldane (1932a) wrote that a mathematical analysis of the effects of selection is necessary and valuable. He wrote: "Many statements which are constantly made, e.g. 'Natural selection cannot account for the origin of a highly complex character,' will not bear analysis. The conclusions drawn by common sense on this topic are often very doubtful." (p. 125) Unaided common sense, according to Haldane (1932a), may indicate an equilibrium, but rarely, if ever, tells us whether it is stable. Haldane stated in his conclusion that if he was only proving the obvious, it is because the obvious was worth proving. He wrote: "And if the relative importance of selection and mutation is obvious, it has certainly not always been recognized as such." (p. 1)

Haldane predicted that the permeation of biology by mathematics, which was evident in theoretical population genetics, had only just begun and will continue in the future, and the investigations summarized in *The Causes* represent the beginning of a new branch of applied mathematics.

Does Natural Selection Work? In *The Causes*, Haldane (1932a) still had to answer critics who questioned the existence of natural selection. He wrote:

> Before ... we deal with the theoretical effects of natural selection, it will be well to give a few examples of it, because the statement is still occasionally made that no one has actually observed it at work. It is quite true that the observations so far made are far from adequate. But at least they prove the existence of natural selection as a fact. (p. 47)

Haldane then proceeded to mention several examples of natural selection at work. For instance, he discussed the case of *Streptococcus haemolyticus*, which rapidly loses its virulence for animals when bred in artificial media. This phenomenon was first misunderstood and misinterpreted as an example of Lamarckian inheritance of the effects of use and disuse. However, Todd (1930) found that when grown on agar, *Streptococcus* produced hydrogen peroxide, but a variant appeared occasionally that was less virulent and produced much less peroxide.

The normal type of bacteria, when grown in agar, make enough peroxide to kill themselves, or at least to slow down their growth rate very greatly. However, when parasitic, they are protected by the catalase of their hosts, which destroys hydrogen peroxide. Hence, the glossy and nonvirulent type, which produces less hydrogen peroxide, is the only survivor after a few weeks of culture. However, if a little catalase is added to the medium, the virulent type grows as well as the nonvirulent and can be preserved in culture indefinitely.

Todd's experiments covered some 2,000 bacterial generations, corresponding to about 50,000 years in human evolution. Haldane mentioned that it took Calmette and Guérin (1924) fourteen years, or about 25,000 generations, to convert the bovine tubercle bacillus into a harmless and indeed beneficial organism by growing it on artificial media. Haldane (1932a) concluded: "There is thus no reason to put down such modifications of bacteria to anything but natural selection, acting on the results of mutation." (p. 51)

Industrial Melanism   In his later years, Haldane used to cite Kettlewell's (1956) observations on Lepidoptera in the industrial regions of England as a well-authenticated instance of directed natural selection. Many species of Lepidoptera have become rapidly dark in areas polluted by industrial smoke. From the time they first appeared, it took only sixty years for the melanics to replace the original type almost completely. The best-studied case is the situation involving the peppered moth, *Biston betularia*. The almost black variety carbonaria first appeared near Manchester in 1848. About 100 years later, its frequencies went up to more than 90 percent over wide areas of England where the lichens on tree trunks have been killed by smoke, darkening the bark. It is due to a single dominant gene C. Another less dark variety, insularia, also due to a dominant gene, is found in partly polluted areas. Haldane (1924) showed that the rapid change could not be due to higher mutation rate which would have to be more than 10 percent. Haldane estimated that the mutation rate may well be 1 per million. He concluded that the observed evolution could be explained if the recessives had an average disadvantage of about 30 percent.

In his experiments in the field, Kettlewell released about equal numbers of dark carbonaria and light recessive moths in a highly blackened wood near Birmingham and an unpolluted wood in Dorset. In collaboration with Nikolaas Tinbergen, he obtained films of predation by small birds that ate the more conspicuous type preferentially. In each wood he released about equal numbers of dark and light moths at daybreak. They were marked with paint under their wings, the mark being visible only during their flight.

In each wood, the number of the more conspicuous phenotype caught was about half that of the phenotype that is hidden when at rest in the wood. However, the intensity of selection was not known. Haldane noted that the frequency of carbonaria has been increasing during his life time, in some areas of southern England, at rates between 1 and 3 percent per year, while it remained stationary in other areas. Haldane (1959) noted: "We can only say that natural selection has largely done its work." (p. 129)

But Kettlewell (1956) has shown something more about evolutionary theory. The carbonaria moths collected in the nineteenth century often showed white patches that would be easily noticed by birds. And when heterozygous carbonaria (Cc) was crossed with the original type (cc) from 1900 to 1905, only 47 percent of the moths that emerged were Cc. But in similar families bred from 1953 to 1956 the frequency was 67 percent. Haldane attributed this significant difference to the selection of genes that improve the health of Cc larvae relative to cc, hypothesizing that these larvae must be biochemically different.

Snails Another example, involving natural selection of very weak intensity, cited by Haldane was the study of Sheppard (1951) on polymorphism in the snail *Cepea nemoralis*. Yellow is recessive to various shades of brown and pink. It is less conspicuous to human eyes on a green or a dark background. Yellow snails are rare in beech woods but are in a majority in permanently green grass. One of their predators, the thrush *Turdus ericetorum*, leaves a record of its predation by breaking the snail shells on stones or stumps, especially during the breeding season.

Sheppard released marked snails in a small wood. The wood was at first brown, and yellow snails were picked out. But as it became greener, fewer and fewer were killed. The selective advantage of one type over another was clearly an average, and if the background had been permanently brown or green, there would have been a considerable advantage.

Haldane noted that most supposed examples of natural selection are very hard to detect. Many economically harmful insects have developed resistance to poison. Laboratory research has shown that this is likely to be a Darwinian rather than a Lamarckian effect, although it may be hard to prove that in the field.

Sexual Selection   Darwin defined sexual selection as the effects of the struggle between the individuals of one sex, generally the males, for the possession of the other sex. He considered that sexual selection acted mainly on males and accounted to a large extent for the bright plumage on many male birds, the horns of male deer, and similar features in other species. Such elaborate displays appear to have two different functions: to attract the females and to conquer or intimidate other males. That distinction is not always clear; the songs of male birds certainly, and their brightly colored plumage probably, drive other males away, as well as occasionally attracting females. Examples of sexual selection abound in many species. For more than a century, sexual selection has become one of the most intensively studied fields of evolution. For example, females of two species of stalk–eyed flies, *Cyrtodiopsis dalmanni* and *C. whitei*, have been shown to prefer males possessing large eyestalks (Wilkinson et al. 1998). In *Drosophila subobscura*, Spurway (1949) found it difficult to mate yellow males with black females. Rendel (1945) investigated this further and found that black females actively repelled yellow males, by dodging, kicking, or extruding the ovipositor. However, by breeding for ten generations from those black females, that did accept yellow males, he obtained a race of black females that accepted 82 percent of yellow males. As Darwin had predicted, the choice depends on hereditary characters of both sexes.

Maynard Smith (1958) showed that virgin females of *D. subobscura* accepted only 45 percent of males of two inbred stocks within one hour, but 90 percent of outbred flies. The success of males' mating capacity depends on their "athletic ability" to keep time with the female dances, and hybrid males were found to be superior in that respect. Because they also fertilize a larger fraction of the eggs, this selection is advantageous to the species.

Speciation by Allopolyploidy   Haldane (1959) had drawn attention to the origin of new species by allopolyploidy, a phenomenon unknown to Darwin. When two related species of flowering plants are crossed, the progeny, though often vigorous, is completely sterile both with itself and with the parent species. Such a sterile hybrid may, however, produce a fertile branch. The progeny of self-fertilized flowers on such a branch are fairly uniform and fertile with one another, but they are usually sterile on crossing with either parent species, and if they can be crossed, the progeny are usually sterile. In other words, they behave as a new species. Thus, allopolyploids are polyploids with chromosomes derived from different species. More precisely, it is the result of doubling of chromosome number in an F1 hybrid. Triticale is an example of an allopolyploid, having six chromosome sets, four from wheat (*Triticum turgidum*) and two from rye (*Secale cereale*).

Muntzing (1932) was the first to repeat, under controlled conditions, a major evolutionary step, when he crossed *Galeopsis speciosa* and *G. pubescens*. The majority of flowering plant species may well be allopolyploids, but the condition is quite unusual in animals mainly because it would interfere with the mechanism of sex determination.

A great number of species with other variants of chromosome numbers have been found. Autopolyploidy with three sets (triploids) or four sets (tetraploids) derived from the same species have been found. These are usually sterile. A species derived from allopolyploidy is, of course, subject to natural selection, and unless it is fitter than either parent, it will die out. The surviving ones evolve as the result of natural selection, adapting to the environment. This subject was reviewed by Stebbins (1950).

Novelty   Critics of Darwinism pointed out at that time that no new characters appeared in these populations as the result of selection.

Haldane responded that novelty is brought about by selection as the result of the combination of previously rare characters. He cited the example of *Primula sinensis* where "the combination of several genes may give a result quite unlike the mere summation of their effects one at a time." (p. 110) A dark stem (recessive) is associated with no great change in color of acid-sapped (red and purple) flowers. On the other hand, blue (recessive) flowers, which have a neutral sap, when growing on a dark stem, are mottled. The same recessive dark stem genes, along with genes for a green stem, give plants that will not set seed, even though they produce good pollen. Selection acting on several characters leads not merely to novelty but to novelty of a kind unpredictable with the available scientific knowledge of the period. Elsewhere, Haldane (1932a, p. 31) noted that intraspecific differences originate partly by combinations of those that existed before and new genes arise from time to time by mutation.

Altruism (Beanbag Genetics):   In the appendix to *The Causes of Evolution*, Haldane (1932a) referred to the evolution of "socially valuable but individually disadvantageous characters." (p. 119) He concluded that the spread of an altruistic gene requires that group size be small and endogamy fairly strict. Haldane wrote:"I find it difficult to suppose that many genes for absolute altruism are common in man." By "absolute altruism," he meant an altruism that is not necessarily directed toward one's relatives. Altruism toward direct descendants can be favored by selection, even in a large unstructured population, but he extended it other relatives as well. As a general statement, Haldane wrote, "For in so far as it makes for the survival of one's descendants and near relations, altruistic behavior is a kind of Darwinian fitness, and may be expected to spread as a result of natural selection." (p. 119)

It is interesting to explore the reasons why Haldane did not proceed further to develop the ideas later formulated by Hamilton (1964), speculating that cooperation in animal societies has evolved because of the genetic relationship of their members. I am in agreement with Maynard Smith (1992) that the problem of the evolution of social behavior did not particularly interest Haldane, whose major concern was the analysis of Darwinian evolution in a Mendelian context.

Furthermore, Haldane wrote that a great deal of human conduct that goes under the name "altruism" is simply egoistic from the point of view of natural selection. He concluded: "It is often correlated with well-developed parental behavior-patterns. Moreover, altruism is commonly rewarded by poverty, and in most modern societies the poor breed quicker than the rich." (p. 122)

Measurement of Natural Selection Mayr (1988) considered natural selection as the most important concept of evolutionary biology, and because of its novelty, "no other concept has encountered as much resistance." (p. 95) Selection, both natural and artificial, received much attention since Darwin's exposition of the theory of evolution by natural selection. One of the earliest measurements of selection was made by Weldon (1901) in the terrestrial mollusk *Clausilia laminata*. By devising certain measurements in their shells, Weldon attempted to estimate the differential selection if any, for certain characters. By comparing similar measurements in the shells of a number of young and old individuals, Weldon showed that the selection for a certain character can be detected and its intensity measured.

The characters measured included the mean angular distance from the columellar plane, and the mean length of peripheral radius. Weldon suggested that if the death rate during growth affects individuals with all kinds of spirals to the same extent, then there will be no change in the mean character and variability of the spiral after its first formation. If, on the other hand, individuals with different kinds of spirals die at different rates, then one would expect to see a change either in the mean length or the standard deviation of radii, or both. One can thus arrive at a conclusion regarding the direction and intensity of selection.

Weldon's approach to selection by means of comparing a set of measurements in two successive generations (young and old) laid the foundation for several later studies. Haldane (1959) concluded that most supposed examples of natural selection are very hard to detect. A well-known case, cited by Haldane, was Darwin's interest in the occurrence of wingless insects on oceanic islands. Darwin thought this was mainly due to the fact that the winged forms were being blown out into the ocean. Haldane commented that no one had introduced a winged species on to such an island to see if it passes though a period when some members are wingless. It is further well known that some insects have developed resistance to such chemical compounds as DDT. Haldane concluded that such instances certainly represent Darwinian selection, not a Lamarckian effect.

Haldane believed that Darwin was justified in arguing from artificial to natural selection. But it was a weak argument. Darwin was not aware that there were stabilizing agencies that could prevent the occurrence of evolution, leading to stagnation or extinction.

R.A. Fisher (1890–1962)

Ronald Aylmer Fisher was born in London on February 17, 1890. He was one of a set of twins, the other having died at birth. Fisher was a precocious child, a trait he shared with Haldane and Wright. He was mathematically gifted but suffered from poor eye sight. It has been suggested that his propensity for geometric proofs, which was so hard for others to comprehend, was due to his not being able to work by lamplight and consequently being tutored orally involving no writing materials (Crow 1984). Fisher was a distinguished student of mathematics and physics at Cambridge University but also expressed a keen interest in the new science of genetics. He was idealistic in the possibility of genetic applications to improve human life (e.g., eugenics) and often expounded on that subject to his fellow students (Box 1978).

Fisher went to Caius College Cambridge, graduating in 1912 with a first in mathematics. Fisher's tutor was an astronomer, and his first paper, "On an Absolute Criterion for Fitting Frequency Curves," published while he was still a student, came out of his study of the theory of errors. However, Fisher's hopes were fixed on the biometricians, and he very badly wanted the "absolute criterion" (the future maximum likelihood) to be noticed by them. In "Mendelism and Biometry," an address to an undergraduate society, Fisher (1911) envisaged a synthesis of these contesting research programs in heredity.

Eugenics and Leonard Darwin Fisher's interest in heredity was combined with a commitment to eugenics; he was one of the founders of the University of Cambridge Eugenics Society (Fisher 1915). This led to a friendship with Leonard Darwin, son of Charles and president of the Eugenics Education Society, which published the *Eugenics Review*, for which Fisher would write many pieces. Darwin's support, financial, intellectual, and emotional, was important to Fisher, especially in the early part of his career.

Fisher (1930) dedicated his most important contribution to evolution, *The Genetical Theory of Natural Selection*, to Leonard Darwin. He wrote: "To Major Leonard Darwin, In gratitude for the encouragement, given to the author, during the last fifteen years, by discussing many of the problems dealt with in this book." When the First World War broke out, Fisher was rejected for military service upon graduation because of his poor eyesight. He offered to serve his country in the Second World War in some capacity but was turned down. In contrast to Fisher, Haldane was actively involved in both wars, fighting in the Black Watch battalion in the first one and conducing spectacular and courageous experiments in diving physiology in the second.

Fisher took up several minor jobs, teaching and working on a farm part of the time. At the end of the war, he was in dire straits because of continuing difficulty to find a suitable position. In desperation, he even thought of becoming a farmer, but, fortunately for statistics, he was eventually hired by the Rothamsted Experimental Station. That was where Fisher made his great contributions to statistics.

Fisher's landmark 1918 paper, "The Correlation between Relatives on the Supposition of Mendelian Inheritance," showing that correlations between relatives are consistent with Mendelian theory and laying the foundations for biometrical genetics, was his first significant contribution to population genetics. He wrote it while teaching physics to school boys and farming. Fisher completed his paper in 1916 and submitted it to the Royal Society of London for publication. The reviewers were Pearson and Punnett, two strong opponents in the famous controversy between the Mendelians and biometricians.

It has been said that the only time when they agreed on anything was when they recommended not to publish Fisher's paper in the *Proceedings of the Royal Society of London*. Fisher turned to his friend Leonard Darwin, who arranged its publication in the *Transactions of the Royal Society of Edinburgh*.

Fisher's problems with Pearson had started earlier. Fisher's remarkable ideas in statistics were neither appreciated nor understood by Pearson. Years later, bitterness still lingered (see Box 1978).

He was a schoolmaster when he published his "Frequency Distribution of the Values of the Correlation Coefficient in Samples from an Indefinitely Large Population" (1915) and on "The Correlation between Relatives" (1918). The first established a new era in the exact theory of sampling distributions. The second vindicated Mendelism and biometry, for it showed how Pearson's biometric results could be explained by Mendelian theory.

In 1919 John Russell of Rothamsted Experimental Station hired Fisher on a temporary basis to see if a statistician could do anything with the mass of data accumulated there. Studies in Crop Variation was the first of a stream of studies showing what could be done. There had been some statistical work on agricultural experiments before the war involving "Student" (W.S. Gosset) and Fisher's Cambridge tutor, the astronomer F.J.M. Stratton, but Fisher raised the subject to a new level. Fisher's genetical research at Rothamsted concentrated on evolution, on integrating Mendelian theory with Darwin's theory of natural selection. His first major theoretical paper was "On the Dominance Ratio" (1922a). He collaborated with E.B. Ford on the analysis of selection in wild populations. Fisher left Rothamsted in 1933, after he had developed the analysis of variance as well as a new approach to experimental design.

Fisher's General Theorems   Fisher was interested in general theorems with respect to the course of evolution, applicable regardless of the number of loci or their interactions. The most important of these was his fundamental theorem of natural selection (1930). Fisher (1922a) attempted to derive the stochastic distribution for his function but obtained incorrect results (Wright 1929), which he corrected in 1930. For the case of a semidominant favorable mutation, he gave the exact probability of ultimate fixation that Haldane's $2k$ closely approximated in large populations (p. 15).

The Genetical Theory of Natural Selection Fisher's (1930) ideas on evolution were brought together in *The Genetical Theory of Natural Selection*. He argued that Mendelism with its view of particulate inheritance did not contradict Darwinism but was consistent with it. Fisher's (1930) classic, *The Genetical Theory of Natural Selection*, is generally regarded as his most important contribution to evolution. "The fundamental theorem of natural selection" was defined by Fisher as follows: "The rate of increase in fitness of any organism at any time is equal to its genetic variance in fitness at that time." (p. 37) The major ideas were that selection acts on the additive genetic component of fitness and that fitness increases at a rate equal to the additive genetic variance at that particular moment. The alleles that prevail in a large population are those that are beneficial in the largest number of combinations, as shown through the scrambling process of Mendelian inheritance. Thus, each species increases in fitness but random processes, such as changes in the environment, may interfere in this process. Fisher also considered the effects of selection in generating linkage disequilibrium.

Genetic variance is defined as the additive component of the total genotypic variance. It was tacitly assumed that the species is essentially homogeneous. It is implied that selection is always according to the net effect of each gene and hence there can be no selection among interaction systems. In such a species, interactional effects among unfixed loci are significant only in reducing the rate of progress by reducing the amount of additive variance, although, of course, a particular interaction system, more or less favorable, tends to be built up step by step, because any mutant gene is advantageous only if it fits in with the genome already established.

Fisher's theory was essentially similar to Haldane's in its deterministic character under given external conditions for genes with more than a few representatives in the population. However, both Fisher and Haldane recognized that the fate of a single mutation is decided by a stochastic process (Fisher, 1922b). Fisher compared his theorem to the second law of thermodynamics. Both are properties of populations, or aggregates, true irrespective of the nature of the units they are composed of, and are statistical laws. Each requires the constant increase of a measurable quantity, in the one case the entropy of a physical system and in the other the fitness of a biological population.

Both Fisher and Wright recognized the importance of gene interaction but arrived at sharply opposing conclusions. Fisher regarded dominance and epistasis as impediments that reduce the effectiveness of selection by decreasing the correlation between genotype and phenotype. Wright, on the other hand, regarded dominance and epistasis as the means for evolutionary progress by producing new harmonious gene combinations.

Fisher concluded his chapter on the "Fundamental Theorem of Natural Selection" with these words: "Although it appears impossible to conceive that the detailed action of natural Selection could ever be brought completely within human knowledge, direct observational methods may yet determine the numerical values which condition the survival and progress of particular species." (p. 51) Fisher (1930) introduced what he called the "Malthusian parameter" as a measure of population increase and, by extension, a "reproductive value." It represents a weight to be assigned to each age group in proportion to the contribution of that group to the future population after age stability has been achieved. Thus, it has both an evolutionary and a demographic significance.

Among other genetic contributions, Fisher generalized Haldane's formula, $P = 2s$, for the probability of ultimate fixation of a gene whose heterozygous selective advantage is s. It was further generalized by Malecot (1952) and Kimura (1964).

University College London   In 1933 Fisher succeeded Pearson as Galton Professor of Eugenics and head of the Galton Laboratory at University College, London. Fisher had much greater admiration for Francis Galton than for his disciple. Though Fisher was Pearson's natural successor in both statistics and eugenics, he did not inherit the whole department, because applied statistics was split off and headed by Pearson's son, E.S. Pearson. This structure did not make for harmony and relations between Fisher and members of Pearson's department, especially its leading theorist, Jerzy Neyman, gradually deteriorated.

On the biological side, Fisher set up a unit to study the genetics of blood groups (see Box ch. 13). The unit, which included G.L. Taylor and R.R. Race, did important work on rhesus blood groups. Fisher also had a breeding colony of mice. In 1943 Fisher returned to Cambridge as professor and head of the Department of Genetics. His theory of inbreeding provided a theoretical analysis of the mouse experiments he had been conducting since his London days. In 1958 Fisher challenged Austin Bradford Hill's inference from the association between smoking and lung cancer that the former was an important cause of the latter. He retired officially from Cambridge in 1957 but stayed until 1959. He spent the last three years of his life in Adelaide, Australia as a research fellow at the Commonwealth Scientific and Industrial Research Organization. His contacts in Adelaide were J.H. Bennett and E.J. Cornish. Fisher died in Adelaide on July 29, 1962, and his ashes lie there in St. Peter's Cathedral.

Sewall Wright (1889–1988)

Sewall Wright was born in Melrose, Massachusetts, to Philip Green Wright and Elizabeth Quincy Sewall Wright. His parents were first cousins, an interesting fact because it may have led to his research on inbreeding. As a child, Wright helped his father and brother print and publish an early book of poems by his father's student Carl Sandburg. Sewall Wright was the oldest of three gifted brothers—the others being the aeronautical engineer Theodore Paul Wright and the political scientist Quincy Wright. From an early age Wright displayed keen interest and talent for mathematics and biology. He was educated at Galesburg High School, and studied mathematics at Lombard College in Illinois, where his father taught. He was greatly influenced by Professor of Biology Wilhelmine Entemann Key, one of the first women to receive a Ph.D. in biology. Wright received his Ph.D. from Harvard University in 1915, under the direction of the great mammalian geneticist William Ernest Castle, investigating the inheritance of coat colors in mammals.

From 1915 to 1925 Wright was employed by the Animal Husbandry Division of the U.S. Bureau of Animal Husbandry. His main project was the close inbreeding in artificial selection that resulted in the leading breeds of livestock used in American beef production. He also performed experiments with 80,000 guinea pigs in the study of physiological genetics. Furthermore, he analyzed characters of some 40,000 guinea pigs in twenty-three strains of brother–sister matings against a random bred stock (Wright 1922). The combined research experience of these mammals eventually led Wright to propose the shifting balance theory and the concept of "surfaces of selective value" (Wright 1931, 1968a). Many of his other important contributions to population genetics took shape while engaging in guinea pig breeding program, including his famous inbreeding coefficient as well as the path coefficient method Wright (1968b).

Wright's longest teaching and research career was in the Department of Zoology at the University of Chicago, where he remained on the faculty for thirty years, from 1925 to 1955. After his retirement from the University of Chicago in 1955, Wright moved to the University of Wisconsin in Madison, where he worked closely with the well-known population geneticist James F. Crow. He remained in Madison until his death in 1988.

Shifting Balance Theory   Wright (1931) joined Fisher and Haldane as a founder of theoretical population genetics through his studies of inbreeding, mating systems, and genetic drift. Wright was the inventor/discoverer of the inbreeding coefficient and F-statistics, standard tools in population genetics. He was the chief developer of the mathematical theory of genetic drift (sometimes known as the Sewall Wright effect), which involves cumulative stochastic changes in gene frequencies that arise from random births, deaths, and

Mendelian segregations in reproduction. Wright was convinced that the interaction of genetic drift and the other evolutionary forces was important in the process of adaptation. He described the relationship between genotype or phenotype and fitness as fitness surfaces or fitness landscapes. On these landscapes, mean population fitness was the height, plotted against horizontal axes representing the allele frequencies or the average phenotypes of the population. Natural selection would lead to a population climbing the nearest peak, while random genetic drift would cause random wandering.

Wright's explanation for stasis was that organisms come to occupy adaptive peaks. In order to evolve to another, higher peak, the species would first have to pass through a valley of maladaptive intermediate stages. This could happen by genetic drift if the population is small enough. If a species was divided into small populations, some could find higher peaks. If there was some gene flow between the populations, these adaptations could spread to the rest of the species. This was Wright's shifting balance theory of evolution. There has been much skepticism among evolutionary biologists as to whether these rather delicate conditions hold often in natural populations. Wright had a long standing and bitter debate about this with Fisher, who felt that most populations in nature were too large for these effects of genetic drift to be important.

Wright strongly influenced several students and colleagues. One of them was Jay Lush, who was the most influential figure in introducing quantitative genetics into animal and plant breeding. Wright's statistical method of path analysis, which he invented in 1921 and which was one of the first methods using a graphical model, is still widely used in social science.

While working at the U.S. Department of Agriculture, Wright conducted extensive research on the genetics of guinea pigs, and several of his students, such as Elizabeth Russell, became influential in the development of mammalian genetics. An anecdote about Wright, which is likely to be true, describes a lecture during which Wright absent-mindedly began to erase the blackboard using a guinea pig!

Wright's approach was directed toward ascertaining whether some way might exist in which selection could take advantage of the enormous number of interaction systems provided by a limited number of unfixed loci. Wright hypothesized that this is possible in a large population subdivided into many small, local populations. Such populations should be sufficiently isolated to permit considerable random differentiation of gene frequencies but not so isolated as to prevent gradual diffusion of the more successful interaction systems from their centers of origin.

Wright (1968a) explained that the process involves three phases:

(a) The first phase is stochastic variability of all gene frequencies in each local population about its set of equilibrium frequencies. The sort of selection involved in determining these equilibrium values was assumed to be, in most cases, that directed toward an optimum not far from the current mean of each quantitatively varying character instead of toward fixation of particular genes. Since many genotypes determine nearly the same intermediate grade under multifactorial heredity, selection toward such a grade implies the existence of many selective peaks in the "surface" of selective values, separated by saddles, and at various heights because of secondary effects on other characters (pleiotropy).

(b) The occasional crossing of a saddle between a lower and a higher peak as a result of extreme stochastic variation leads to the second phase, in which selection is directed toward the set of equilibrium frequencies of the higher peak, a process to which Haldane's formulas (and also Fisher's fundamental theorem) apply.

(c) The final phase is the spreading of this superior system by excess population growth and emigration to neighboring populations and ultimately, perhaps, throughout the species as a whole. It was further recognized that local selection in a different direction may occasionally lead to superior general adaptation, replacing the first two phases. This theory is applicable in pure form only if there is a favorable state of subdivision of the species and external conditions are stable over a period of many generations.

H.J. Muller

In addition to the trio of pioneers who founded the mathematical theory of evolution, another name, H.J. Muller, must be added. Although Muller's contributions to evolutionary biology were entirely nonmathematical, through his ingenious experiments in *Drosophila* genetics and his impressive insights, Muller contributed to the genetic basis of evolution. This subject was adequately summarized by Crow (1992) and Carlson (1981). Of special interest is Muller's view of gene primacy, an important forerunner of the "selfish" gene concept. Muller was one of those who believed that the first meaningful form of life might have been a primitive gene and that the property of making errors and copying these errors at the fundamental level of life, that is to say, represents a system that is capable of evolving by means of natural selection.

A Synthesis

It has been suggested that a synthesis between classical genetics and population genetics was critically responsible for the emergence of modern evolutionary theory in the 1920s (Sarkar 2004). Haldane's (1932a) book differed from the works of Fisher (1930) and Wright (1931) in being a much broader discussion of concerns beyond population genetics. Furthermore, the appendix of Haldane's *The Causes* covered almost all the mathematical models of population genetics that were known until then. The text of *The Causes* attempted a comprehensive summary of all known mechanisms of evolution, from the point of view of both classical genetics and cell biology. In that sense, *The Causes* represented, more accurately, evolutionary biology rather than population genetics. The presentations of Fisher and Wright, especially Wright, were much narrower in scope from a broader biological point of view than Haldane's.

In his discussion in *The Causes*, Haldane (1932a) touched upon several other issues, including the ethical consequences of evolution, the possibility of divine guidance in evolution, evolution of altruism, and the evolution of mind, among others.

3

Time Line: J.B.S. Haldane (1892–1964)

1892, November 5	Born in Oxford, England, the son of John Scott Haldane, physiologist at New College, Oxford, and Louisa Kathleen Haldane.
1895	Injured in an accident and questions the attending physician whether the blood from his wound contained hemoglobin or carboxyhemoglobin.
	Attended Oxford Preparatory School (now Dragon School, Oxford); won a scholarship to go to Eton where he excelled in Latin, Greek, German, French, history, chemistry, physics, and biology. Julian Huxley was his senior, and the two became friends, establishing a lifelong friendship which also included Julian's brother Aldous and Haldane's sister Naomi.

1900   Started helping his father in recording results of physiological experiments. Accompanied father to a lecture by A.D. Darbishire on Mendel's experiments at the Oxford University Junior Scientific Club which created a lasting interest in genetics,

1906   Assisted his father in physiological experiments by going down mines to collect air samples and diving to test equipment.

1911   Went to New College on a mathematical scholarship; attended E.S. Goodrich's final honors course in zoology, which created a lasting interest in evolutionary biology and genetics.

At a seminar organized by Goodrich, Haldane reported the first case of linkage in mice.

1912   Published his first scientific paper, in the *Journal of Physiology*, in collaboration with his father and C.G. Douglas on the laws of combination of hemoglobin with carbon monoxide and oxygen.

1915   Published a paper (with his sister, Naomi, and A.D. Sprunt) in the *Journal of Genetics*, reporting the first case of linkage in vertebrates.

World war I, Joined the Black Watch Battalion, and fought in the trenches of France.

1918   Sent to Simla, India, to recuperate from wounds received in explosions in Mesopotamia (Iraq); fell in love with India but determined to return when he could associate with Indians "on a footing of equality."

1919   Invented the first mapping function and the unit of map distance, the centimorgan (cM).

Appointed Fellow in Physiology at New College, Oxford.

1920   Described the gene as a self-reproducing nucleoprotein molecule.

1922   Proposed "Haldane's rule" (Haldane 1922). "When in the F1 offspring of two different animal races one sex is absent, rare, or sterile, that sex is the heterozygous sex." This rule has been shown to be true for a great number of interspecific crosses across several phyla. The fact that hybrid sterility and inviability can evolve due to Haldane's rule in such a vast array of different organisms is quite impressive. However, the actual explanation of this phenomenon is still undecided.

1923   Lectured before the Heretics at Cambridge University, met C.K. Ogden, who was a "scout" for Kegan Paul who published Haldane's lecture under the title, *Daedalus, Or Science and the Future,* a futuristic essay, influenced by the books of H.G. Wells, speculating important future developments in science and technology and especially noted for predictions in molecular and reproductive biology, human cloning, raising ethical and moral issues (Appointed Sir William Dunn Reader in Biochemistry under Prof. F.G. Hopkins (later Nobel laureate and President of the Royal Society of London).

1924   Published the first paper in his series, A Mathematical Theory of Natural and Artificial

Selection." (one of the foundations of population genetics),

Interviewed by Mrs. Charlotte Burghes, a reporter for London's *Daily* Express and his future wife.

1925   Cited as a co-respondent in the divorce case involving Charlotte Burghes; appeared before the *Sex Viri* court, a panel of six men chosen from the faculty; dismissed by Cambridge University on moral grounds but was reinstated on appeal with support from his father and Hopkins,

With G.E. Briggs, derived the basic law of steady-state kinetics that is still used for treating enzymatic catalysis. Haldane foresaw the value of specific enzymes in synthetic chemistry and industrial applications.

1926   Married Charlotte Burghes; moved to Roebuck House in Cambridge.

1927   *Possible Worlds and Other Essays* includes Haldane's most famous essay, "On Being the Right Size."

Investigated the role of carbon monoxide as a tissue poison; Haldane was the subject-investigator.

Joined John Innes Horticultural Institution as Officer-in-Charge of Genetical Investigations, a part-time position; guided C.D. Darlington's research in cytogenetics.

1929   Proposed his theory of the origin of life, independent of A.I. Oparin in the Soviet

Union, suggesting that primitive life arose in an anaerobic prebiotic world, facilitated by the synthesizing action of ultraviolet light on a mixture of gases in the ocean, resulting in the "consistency of hot dilute soup," so that any organism that appeared would have abundant food.

1930   *Enzymes*, first book on enzymes in English language.

1932   Elected Fellow of the Royal Society of London.

The Causes of Evolution a popular exposition of population genetics, including a summary of his papers on the mathematical theory of natural selection, first estimation of a human mutation rate, a quantitative treatment of genes for altruism in human populations, and the ethical and moral implications of Darwinian evolution. Also *The Inequality of Man and Other Essays*, an outstanding collection of his popular essays and a science fiction story ("The Gold Makers").

An important paper on the role of gene action at different stages of the life-cycle of organisms; "The time of action of genes, and its bearing on some evolutionary problems."

Formal genetics of man (A method for investigating recessive characters, using Maximum Likelihood Method)—one of the foundations of human genetics Haldane (1932b).

1933   Appointed Professor of Genetics, University College, London.

1934	"Quantum Mechanics as a Basis for Philosophy", stimulated Norbert Weiner's idea of cybernetics.
	Fact and Faith (London: Watts).
1935	*Science and the Supernatural*, (London: Kessinger Publishing Co) (with Arnold Lunn;), based on an exchange with Lunn on science versus religion; and *Science and Well-being* (London: Kegan Paul).
	Estimation of human mutation rate (with Lionel Penrose)
1936, March 14	Father John Scott Haldane died; joined the Spanish civil war, traveled to Spain.
1937	*My Friend, Mr. Leaky* (London: Cresset Press), a storybook for children.
	First human gene map (with Julia Bell): linkage estimation between the genes for hemophilia and color-blindness.
	Chairman, Editorial Board, *Daily Worker*.
	"The Effect of Variation on Fitness" the basis for the theory of genetic loads, and later estimates of the genetic effects of ionizing radiation resulting from the open air testing of nuclear bombs.
1938	*A.R.P.* (London: Victor Gollancz); campaigned actively to promote a sound policy of air raid protection. Subsequent events during the war proved he was correct. Two important books: *Heredity and Politics* and *The Marxist Philosophy and the Sciences*; Also 38th Robert Boyle Lecture, Oxford, *The Chemistry of the Individual*.

1939   Invited by the British Admiralty to investigate the sinking of the submarine H.M.S. *Thetis* off the coast of Liverpool, with the loss of 99 lives, including many civilians; tested conditions of escape from a steel chamber that simulated the circumstances in the escape chamber of the *Thetis*.

1940   *Science in Peace and War, Science in Everyday Life* and *Keeping Cool and Other Essays*

1941   Tasted oxygen at high pressures; human physiology under high pressure and very cold temperatures while breathing nitrogen and carbon dioxide (Haldane was the subject-investigator).

              New Paths in Genetics, introducing "cis" and "trans" to replace "coupling" and "repulsion" phases in linkage which were introduced by William Bateson in 1906.

1942   "The Selective Elimination of Silver Foxes in Eastern Canada"

1944   "Mutation and the Rhesus Reaction"

              Studied the role of helium in deep sea diving, and the role of radioactivity in the origin of life in Milne's cosmology.

1945   With the rise of Trofim Lysenko in Soviet science, attacks on Mendelian geneticists, and destruction of textbooks in genetics, Haldane was caught in a political crisis between his Marxist sympathies and his loyalty to the science of genetics.

1945, January	Sent a letter to Lysenko requesting copies of his papers with his experimental results that led to his "views on genetics" and his conclusions, but did not receive a reply from Lysenko
1946	*A Banned Broadcast and Other Essays.* Analyzed the interaction of nature and nurture (heredity vs. environment).
1947	*Science Advances*, London: Allen & Unwin and *What Is Life,* New York: Boni & Gaer?
	Studied the genetic effects of atomic bomb explosions; X-chromosome mapping (with C.A.B. Smith), which improved the earlier map drawn by Julia Bell and Haldane in 1937,
1948	Delivered the Royal Society's Croonian Lecture, "The Formal Genetics of Man," an important step in the consolidation of "human genetics" as a distinct discipline.
	Attended the Eighth International Congress of Genetics, Stockholm,
	Estimated the rate of mutation of human genes.
	Haldane made the important suggestion that the heterozygous carriers for thalassemia may possess greater immunity to malarial infection which could account for the higher frequencies of thalassemia in malarial regions (in contrast to the higher mujtation rate that was erroneously postulated by J.V. Neel).
1949	A seminal essay, "Disease and evolution," initiated much epidemiologic research, Haldane, J.B.S. (1949b). Developed a general theory of

evolution in which resistance to infectious disease plays a crucial role. Parasitism may have been an important factor in speciation.

1951   Published *Everything has a History*, London: Allen & Unwin; Invited by Arthur C. Clarke to address the British Interplanetary Society in London; gave a lecture titled "Life in Space, Space Ships and Other Planets." This was the beginning of a long friendship with Clarke.

1952   The Haldanes visited India at the invitation of the Indian Science Congress Association and Prof. P.C. Mahalanobis, Director of the Indian Statistical Institute in Calcutta.

1953   Delivered lecture titled "Genetics of Population Structure" at the International Biological Union Symposium, University of Pavia, Italy. The address was noted for its many quotes from the classics, an impressive display of classical knowledge.

1954   "The Statics of Evolution," in Haldane, JBS (1954) *Evolution As a Process*, which discussed both the statics and the dynamics of evolution; and *The Biochemistry of Genetics*, London: Allen & Unwin.

1956   Huxley Memorial lecture and Medal of the Royal Anthropological Institute, London.

1957   Discussed a neglected aspect of evolution in an important paper, "The Cost of Natural Selection", which showed that the cost incurred by the species, in deaths, during the process of gene substitution, could be as high as thirty times the population number in a single generation. Kimura (1968) used Haldane's estimate as a justification for the "neutral theory of evolution."

1957, May	Delivered the Karl Pearson Centenary Lecture, University College, London.
1957, July	The Haldanes moved to India to work at the Indian Statistical Institute in Calcutta.
1957, December	Delivered the Vallabhbhai Patel Memorial Lectures, titled "Unity and Diversity of Life," on All India Radio, New Delhi. The author joins Haldane's team.
1958	Chief Guest of the Assam Science Society, Gauhati, Assam.
1958, January	Delivered "The Present Position of Darwinism," at the 45th Indian Science Congress Association, Madras, India.
1959	Delivered Yellapragada Subba Rao Memorial lecture "Genetics in Relation to Medicine" at the Academy of Medical Sciences, Hyderabad, India.
1959–1961	Member of the Indian Government's University Grants Commission committee to review teaching and research.
1960, January	Attended the Indian Science Congress meeting, Mumbai.
1960, December	Chief Guest of the Ceylon Association for the Advancement of Science, hosted by Governor General of Ceylon Sir Oliver Goonetilleke; met with Arthur C. Clarke.

1961	Presented research on human relationships at the Second International Congress of Human Genetics, Rome, Italy, and
	President of the International Conference on Human Population Genetics, Hebrew University, Jerusalem, Israel.
1962	Moved from Calcutta to Bhubaneswar to found the Genetics and Biometry Laboratory with the support of the Orissa state government, India.
1963	Published papers in theoretical population genetics in collaboration with S.D. Jayakar.
	"Biological Possibilities for the Human Species in the Next Ten Thousand Years" Lecture at a CIBA Foundation Symposium in London,
	Delivered "The Implications of Genetics for Human Society" at the 11th International Congress of Genetics, The Hague, which introduced the term "clone."
1963, October–November	Visited Europe and the United States; lectured at Rockefeller University, University of Rochester, and the University of Wisconsin; participated with A.I. Oparin in a symposium on the origin of life at the University of Florida.
1963–1964	Underwent surgery for rectal cancer at University College Hospital, London, and returned to Bhubaneswar, India.
1964	In response to Ernst Mayr's challenge questioning the significance of the theoretical

contributions of Fisher, Haldane, and Wright, Haldane wrote "A Defense of Beanbag Genetics", a spirited defense of mathematical population genetics.

Died in Bhubaneswar, India, on December 1, 1964

Haldane's body sent to the Medical College at Kakinada, South India, as desired in his will.

1976   Last posthumous publication *The Man with two Memories* (London: Merlin Press)

1977   Mrs. Haldane (Dr. Helen Spurway) died in Hyderabad, India.

4

Time Line: Ernst Walter Mayr (1904–2005)

1904, July 5	Born in Kempten, Bavaria, Germany, son of Otto Mayr (1867–1917), jurist in the Bavarian court system, and Helen Pusinelli Mayr (1870–1952); displayed childhood interests in natural history, especially bird-watching.
1917, July 1	Father died; moved with mother and two brothers to Dresden; attended the Staatsgymnasium.
1922, February	Passed the Abitur and graduated from the Gymnasium; received a pair of binoculars as a present from his mother; spent the next several weeks on daily bird-watching trips.
1922, April	Joined the Saxony Ornithological Association.

1923, March 23	Observed a pair of red-crested pochards (*Netta ruffina*) on the Frauenteich, one of the lakes of Moritzberg; this species had not been observed in Saxony since 1845. He was unsuccessful in showing these birds to other members of the Saxony Ornithological Association; Dr. Raimund Schelcher provided a letter of introduction to Dr. Erwin Stresemann, newly appointed curator of birds in the Natural History Museum, Berlin.
1923, April	Stopped in Berlin on his way to the University of Greifswald to begin his medical studies; convinces Stresemann of his observations of the red-crested pochards; Stresemann published this record (with the incorrect date of 1922, rather than 1923) and invited Mayr to work at the museum during term holidays, which he did and was captivated ("It was as if someone had given me the key to heaven.").
	The choice of medical studies at the University of Greifswald was not based on any excellence of this university, but on its location in one of the ornithologically interesting areas of Germany. Mayr spent considerable time observing birds and thinking about problems in avian biology.
1925, February	Passed all preclinical subjects and obtained the degree of Candidate of Medicine, which would allow him to complete his medical studies in the future if desired;

spent the last semester at the University of Greifswald as a student of zoology.

1925, October	Started his studies at the University of Berlin; thesis topic was the spread of the serin finch (*Serinus canaria canaria*) in Europe, which was published in 1926.
1926, June 26	Passed his Ph.D. examination and was appointed an assistant in the museum on July 1. His major task was in the library, where he developed with W. Meise a catalog of the journals in the museum library, published in 1929 and used until 1992, when a computerized catalog was developed for the museum library; his major systematic work was a revision of the snow finches (*Montifringilla* and *Leucosticte*) published in 1927.
1928–1930	Participated in expeditions to New Guinea and the Solomon Islands. As an incentive to change his studies to zoology, Stresemann promised Mayr that he would arrange an expedition to some exotic country. This was finally arranged with a joint expedition to New Guinea under the auspices of Walter Rothschild (Tring Museum), L. Sanford (American Museum of Natural History [AMNH]), and the Berlin Museum. Mayr left Berlin on February 4, 1928, and on March 4 arrived in Jakarta, Indonesia, where he received equipment, several assistants, and further instructions in tropical collecting.

1928, April 5	Arrived in New Guinea; this expedition had eventually three independent parts: Dutch New Guinea, Papua New Guinea, and the Solomon Islands, the last was part of the AMNH Whitney South Sea Expedition (WSSE).
1929, May 10	Received invitation to join the WSSE; accepted and joined the expedition on the schooner *France* on July 3 in Samarai, Papua New Guinea.
1930, February 17	Left the WSSE at Tulagi, Solomon Island; departed for Marseilles on March 5, arriving in France at the end of April and Berlin shortly thereafter. Attended the Seventh International Ornithological Congress in Amsterdam in June, where he met Frank Chapman (ornithology, AMNH). He also attended the 1934 and 1938 congresses, and the next congress in 1954 in Basel after several years.
1930, October	Received invitation for a year's temporary position at the AMNH
1931, January	Departed Germany and reached New York on January 19; reported for duty at the AMNH on January 20. He stayed first at the International House just opposite Grant's Tomb and then shared an apartment with two other men on Tiemann Place, very close to the International House. Mayr's invitation to the AMNH was the decision of Dr. Leonard C. Sanford (AMNH trustee and strong supporter of its ornithology department), who had

obtained funds for the WSSE, Mayr's salary during his entire stay at the AMNH, purchase of the Rothschild collection, and the building housing the ornithology department.

1931, March 31

Published his first paper on work done at the AMNH; published six more papers that year on his analyses of birds from the WSSE. Most papers during the 1930s dealt with his studies on the bird collection of the AMNH. Museum authorities disallowed any overseas fieldwork, keeping Mayr chained (so to speak) to his desk with the comment that enough unstudied material of birds from earlier expeditions already exists in the cabinets; Dr. Leonard C. Sanford befriended Mayr and acted as a surrogate father.

1931–onward

Mayr joined the Linnaean Society of New York (a birding club) and was active in this organization as, among other things, editor of their proceedings; became honorary member of the Bronx County Bird Club, where he directed members in field studies of birds.

1932

The AMNH purchased the Rothschild collection (Tring, England) and appointed Mayr the first and only Whitney-Rothschild curator of birds, ensuring he would remain at the AMNH. Mayr then resigned his position at the Berlin Museum and became responsible for unpacking the Rothschild collection and integrating it with the rest of the bird collection of the museum. He was also in charge of developing the public exhibitions occupying the first and second floors of the Whitney Wing, especially for first floor's Biology of Birds Hall. During his tenure at the AMNH, Mayr worked on the systematics and distribution of birds, establishing a strong empirical foundation for his studies in evolutionary biology and in the history and philosophy of biology.

1933   Published "Die Vogelwelt Polynesiens," in which he first outlined the ideas of island biogeography.

1934   Developed kidney illness and had left kidney removed.

1935, May 4   Married Margarete (Gretel) Simon in Freiburg im Breisgau; set up housekeeping in Inward Park at the northern tip of Manhattan. The couple had two daughters, Christa and Susanne.

1936, October   Met Theodosius Dobzhansky when he presented his lectures on population genetics and evolution at Columbia University (the basis of Dobzhansky's 1937 book *Genetics and the Origin of Species* [New York: Columbia University Press]) and demonstrated his work at the AMNH on geographic variation and speciation in birds; Mayr and Dobzhansky became close lifelong friends.

1937, April	Moved his family to Tenafly, N.J., where he spent much time doing field studies of birds in Bergen County after the move—he published on some of these studies but did not incorporate ecological work into a major part of his evolutionary analyses.
1939, December 28	Lecture on speciation in birds at the joint symposium of the American Society of Naturalists and the Genetics Society of America; was invited immediately by L.C. Dunn to present the 1941 Jesup Lectures on "Speciation and Evolution" at Columbia University together with Edgar Anderson (botanist).
1940s	Active in the work of the Committee on Common Problems of Genetics, Paleontology and Systematics (National Research Council, Washington, D.C.); was a major figure in the founding of the Society for the Study of Evolution, being Secretary (1946), founding editor of its journal *Evolution* (1947–1949), and president (1950); invited J.B.S. Haldane to join the editorial board.
1940	Published two papers on the distribution and history of Polynesian birds, restating the ideas of island biogeography.
1941	Published his first book, *List of New Guinea Birds*. New York: American Museum of National History.

1941, March	Presented four of the eight Jesup Lectures at Columbia University
1942	Published *Systematics and the Origin of Species* (New York: Columbia University Press) based on his Jesup Lectures.
1945	Published *Birds of the Southwest Pacific.* New York: The Macmillian Co.
1946	Published *Birds of the Philippines.* New York: The Macmillian Co.
1948	Took part in the 1948 Princeton Conference and coedited the 1949 volume *Genetics, Paleontology and Evolution*; this conference is usually taken as the end of the modern evolutionary synthesis.
1949	Was visiting professor at the University of Minnesota, Minneapolis.
	Coedited *Ornithologie als biologishe Wissenschaft*, festschrift for the 60th birthday of Erwin Stresemann (with Ernst Schuz) Heidelberg: Carl Winter.
1951, December	Ernst and Gretel receive their U.S. citizenship (their children were born U.S. citizens); they had not left the United States using their German passports since before the start of the Second World War. They traveled to Europe in 1952.
1952	Was visiting professor at the University of Washington, Seattle; taught a course on evolutionary theory during the same term that R. Goldschmidt also taught a course in evolution.

Received inquiry about his possible interest in the Alexander Agassiz Professorship
of Zoology at the Museum of Comparative
Zoology (MCZ), Harvard University, and
a definite invitation early in 1953. Mayr
was interested in a more academic position, teaching courses and having graduate
students and a shorter commute (compared to the hour trip between Tenafly and
the AMNH); since the death of Dr. L.C.
Sanford on December 7, 1950, Mayr had
felt "free" to leave the AMNH. He
remained a member of the Department of
Ornithology at AMNH and was elected a
Trustee of the museum for two terms.

1953   Published *Methods and Principles of Systematic Zoology*, (New York: MacGraw-Hill)
and two further editions in 1969 and 1991.

1953, July   Resigned from the AMNH and assumed
his duties at the MCZ, Harvard University;
retired in 1975 and remained associated
with the museum and Harvard University
until his death in February 2005—a total
period of 52 years.

1953, August   Took an active part in the Colloquium on
Zoological Nomenclature, Copenhagen.

1953–1959   Vice president, American Ornithologists'
Union, 1953–1956; president, 1956–1959.

1953–1986   Major organizer and editor of volumes
8–15 and revised volume 1 of the *Check-
list of the Birds of the World*, (Cambridge:
Harvard University Press) a volunteer task
he assumed upon arriving at the MCZ.

1954	Elected to the National Academy of Sciences.
	Purchased a summer place ("The Farm") in Wilton, N.H. (southern part of the state), where the family spent weekends from late spring to the end of November and summers when they were not traveling.
1954–1976	Served as commissioner of the International Commission of Zoological Nomenclature.
1957	Received honorary Ph.D. from Uppsala University (first of 17 such degrees).
	Edited *The Species Problem*. Washington, D.C. American Association for the Advancement of Science.
1958	Served as vice president of the Eleventh International Zoological Congress.
1959	Moved to a house at 11 Chauncy Street, Cambridge, after renting an apartment for six years.
1960–1961, winter–spring	Sabbatical period (only one): trip to Australia and India, his first visit to exotic areas since returning from WSSE in 1930; met J.B.S. Haldane and the present author at the Indian Statistical Institute in Calcutta, India; went bird-watching in Bhubaneswar, Orissa, with the Haldanes and the present author.
1961–1970	Served as director of the MCZ; obtained funds for its new experimental wing and the 700-acre field station in Concord, Mass.

| 1962 | Served as president of the Thirteenth International Ornithological Congress, Ithaca, N.Y. |

1970   Published *Populations, Species and Evolution* (Cambridge, Mass.: Harvard University Press), which was a greatly revised and shortened edition of *Animal Species and Evolution* (Cambridge, Mass.: Harvard University Press, Mayr, E. (1963).

1975   Retired and became Alexander Agassiz Professor of Zoology Emeritus.

1976   Published *Evolution and the Diversity of Life* (Cambridge, Mass.: Harvard University Press, 1963), comprising mainly edited essays of earlier papers.

1980   Coedited *The Evolutionary Synthesis* (Cambridge, Mass.: Harvard University Press) with W. Provine, comprising published papers from a workshop on the history of the evolutionary synthesis organized by Mayr.

1982   Published *The Growth of Biological Thought* (Cambridge, Mass.: Harvard University Press), a history of evolutionary biology and related subjects.

1983   Was awarded the Balzan Prize by the Balzan Foundation, Milan, Italy, the first time the prize had been awarded to a zoologist.

1988   Published *Toward a New Philosophy of Biology* (Cambridge, Mass.: Harvard

University Press), comprising mainly edited essays of earlier papers.

1990, August

Gretel Mayr died; during the following winters he spent several of the cold months in warmer climates, first in Panama and then at Rollins College in Orlando, Fla.

1991

One Long Argument: Charles Darwin and the Genesis of Modern Evolutionary Thought (Cambridge, Mass.: Harvard University Press).

1994

The MCZ dedicated the Ernst Mayr Library.

1994, summer

Attended a symposium at the meeting of the International Society for the History, Philosophy and Sociality of Biology in honor of his 90th birthday.

1994, November

Was awarded the International Prize for Biology (Japan Prize).

1995

Contributed a chapter to *Haldane's Daedalus Revisited* (Dronamraju 1995).

1997

Moved from 11 Chauncy Street to a retirement home in Carlton-Willard Village, Bedford, Mass., close to his daughter Susie.

Published *This Is Biology: The Science of the Living World* (Cambridge: Harvard University Press).

1999

Received the Crafoord Prize, Stockholm.

2001

Published *What Evolution Is*, (New York: Basic Books) a general, beginning text on evolutionary theory, as well as *The Birds of*

	Northern Melanesia (with Jared Diamond), (New York: Oxford University Press), his final detailed analysis of his ideas on island biogeography.
2004	Published *What Makes Biology Unique* (Cambridge: Cambridge University Press), comprising mainly edited essays of earlier papers.
2004, May 10	Attended a symposium organized by the MCZ in honor of Mayr's 100th birthday.
2004, August	Attended a symposium at the annual meeting of the American Ornithologists' Union in honor of his 100th birthday, including a video interview of Ernst Mayr made on November 8, 2003, which was published in 2005 after Mayr's death.
2004, fall	Diagnosed with secondary cancerous lesions of the liver; admitted to the nursing wing of the Carleton-Willard; took physical therapy in the hopes of returning to his apartment, but never did.
2004, December	Received the Gold Medal from the Accademia dei Lincei, Rome, Italy.
2005, February 3	Died at Carleton-Willard Village, Bedford.
2005, July 5	Ashes of Ernst Mayr were scattered along the glacial esker on the shore of Burton Pond, The Farm (Wilton, N.H.), to join those of his wife, Gretel.

5

J.B.S. Haldane and Evolutionary Biology

John Burdon Sanderson Haldane (1892–1964), or J.B.S., as he was widely known, was a polymath who never earned a higher degree in science, yet his contributions to physiology, biochemistry, genetics, biometry, statistics, cosmology, and other sciences have transformed science quite significantly. Much of his scientific knowledge was self-taught.

Haldane's scientific contributions can be classified into three main categories. First, his mathematical theory of evolution is generally regarded as his most important contribution to science. Second, his intellectual contributions involving several sciences have helped to establish new scientific disciplines by cross-pollinating ideas and concepts from diverse disciplines such as population genetics, biochemical genetics and sociobiology. Third, his highly skilled scientific essays that were published in the popular newspapers and magazines during the years 1923–1964 were a true contribution to popular science writing and science education.

Haldane also made important contributions to human physiology, which involved conducting painful experiments upon himself and his associates while testing the effects of breathing various gaseous mixtures under extremes of temperature and pressure (see Behnke and Brauer 1968). Some of this work was done when he was investigating escape from submarines under water for the British Admiralty. In biochemistry, in a paper with G.E. Briggs, Haldane derived the basic law of steady-state kinetics that is still used for treating enzymatic catalysis (Briggs and Haldane 1925).

Family Background and Childhood

J.B.S. Haldane was born in Oxford, England, in 1892. He was the only son of the eminent Oxford University physiologist John Scott Haldane (1864–1936), who groomed his son from an early age for a scientific career. His mother, Louisa Kathleen Trotter, was a member of a comfortably equipped south Scottish family. An active member of the Victoria League and an empire loyalist, she was clearly at odds with the left-leaning Oxford society of JBS's childhood.

Haldane's ancestors were noted for their intellectual distinction in science, politics, and literature. Several of them were elected to the Royal Society, as was Haldane. Haldane's uncle was Lord Haldane, who served as Chancellor of the Exchequer and later Minister for War during the First World War. His great uncle Burdon Sanderson was the first Waynflete Professor of Physiology at Oxford University. His sister, Naomi Mitchison, who was married to the Labor Member of Parliament Richard Gilbert Mitchison, was a distinguished and prolific writer of both fiction and nonfiction. Naomi's three sons, Dennis, Murdoch and Avrion are distinguished scientists in biology and medicine.

J.B.S. Haldane's intellectual abilities became evident from an early age. Even as a toddler he used to watch his father conducting physiological experiments in the basement-laboratory. He could read by his third birthday. And before his next birthday he is claimed to have asked, on looking at the blood from his cut forehead: "Is it oxyhemoglobin or carboxyhemoglobin?" His mathematical prowess also soon became evident. Before he was ten he was assisting his father in routine calculations and, on one occasion, while on an expedition he calculated a set of log tables when his father asked him to do so.

It seemed preordained that the younger Haldane would grow up to be a scientist in his father's footsteps. He was sent to the Oxford Preparatory School (now the Dragon School), where he excelled in Latin, arithmetic, and geometry, winning the first scholarship to Eton.

At Eton, Haldane chose a broad selection of subjects for his studies, including Latin, Greek, German, French, history, chemistry, physics, and biology. Haldane later wrote that he was not at all happy at Eton, although he received an intellectually sound education. His brilliance simply alienated him from most of the students as well as the faculty, and he was much tormented by the senior boys. But Haldane eventually survived the ordeal, becoming the captain of the school, a position of great importance at Eton. J.B.S. later wrote that he could read Latin, Greek, French, and German and learned enough chemistry and biology to do unaided research. Winning a mathematical scholarship, Haldane went up to Oxford, fully prepared to meet new challenges. He received formal education at Oxford University, graduating in classics in 1915, but never earned a degree in science. His higher education was interrupted when he joined the Blackwatch Battalion, serving with distinction during the First World War (1915–1919).

After the war, he was awarded a fellowship in physiology at New College, Oxford, a position he occupied until 1923, when he accepted the Sir William Dunn Readership in Biochemistry at Cambridge University. Haldane's Cambridge years, 1923–1933, were most important in his professional and personal growth, when he first came to world's attention as the famous and eccentric "J.B.S." It was also during those years that Haldane became his own "guinea pig" when he conducted painful physiological experiments while testing the toxic effects of various gases such as carbon monoxide under various experimental conditions. Later, he conducted similar experiments in the years preceding and during the Second World War in response to a request by the Admiralty to investigate the accidental loss of a submarine in which many sailors lost their lives.

From 1933 onward, he was the famous "Prof." at "UCL" (University College London), first in genetics and later as the Weldon Professor of Biometry (1937–1957). They were tumultuous years when his involvement in endless controversies was fodder for the daily press. Very much larger than life, it seemed at times that Haldane's activities were mostly disconnected with his chair at UCL. Yet, to say that Haldane neglected his academic responsibilities would not be accurate. He continued his teaching and research at a breakneck pace, publishing numerous books and research papers of the highest quality. He was very much in demand as the preferred speaker at various academic and public functions as well as the major contributor to numerous books and other publications. His popularity as a public speaker enhanced even further when his capacity to speak on any topic at a moment's notice became well known. He clearly enjoyed his increasing fame and wide contacts with the public. His fame was further augmented by the numerous popular essays which he contributed to the daily press, occasional radio broadcasts, and his frequent press interviews on any topic under the sun.

One major controversy was his public protest against Anglo-French attack on Egypt when the Suez Canal was nationalized by Nasser in 1956. Although he was quite unhappy for several years with his life in Great Britain and British politics, the Suez crisis acted as the trigger that propelled Haldane to leave Great Britain and migrate to India. He moved to India permanently in 1957 to accept the research professorship at the Indian Statistical Institute in Calcutta. His wife, Dr. Helen Spurway, was also offered a research position at that institute. Haldane became an Indian citizen in 1961 and died of cancer in India in 1964.

From about 1937 until 1950, Haldane embraced Marxism and wrote a great deal on the Marxist view of science and social problems. Many of his outstanding popular scientific essays written during those years appeared in the *Daily Worker* of London (Dronamraju 2009). However, his Marxist philosophy did not appear to have any significant impact on his scientific work. Over the years, Haldane's scientific colleagues and associates came to disregard his political views when evaluating his great contributions to science.

Scientific Work

Haldane's earliest research publications were in respiratory physiology in collaboration with his father, but his most important paper from his early research was in genetics. In collaboration with his sister, Naomi, and C.G. Douglas, Haldane reported the first case of "linkage" in vertebrates in 1915. Returning from the war in 1919, he published another important paper that pioneered the concept of "mapping function" as well as the suggestion that the centimorgan, or cM, should be used as a unit of map distance in gene mapping. Another important paper that appeared in 1922 formulated the rule (later called Haldane's rule) that "when in the offspring of two different animal races one sex is absent, rare, or sterile, that sex is the *heterozygous* (heterogametic) sex." (Haldane 1922, p. 101). This rule has withstood the test of time for more than eighty-five years and is now applicable to many species at all levels of the animal kingdom.

Daedalus   Haldane's little book *Daedalus; or, Science and the Future,* which was published in 1923, stood apart from his other books and research publications. Indeed, it was unique and remains so even today among all his publications. *Daedalus* was a highly speculative essay that contained bold predictions regarding the future of the human species. It touched upon such speculative subjects as *in vitro* fertilization, genetic manipulation of the fetus, large-scale breeding of individuals with exceptional qualities, and the new ethical dilemmas created by the new technologies. Haldane's prescient discussion of *in vitro* fertilization and selective breeding of children with special qualities drew much attention and notoriety in the popular press, causing much embarrassment to his parents. Any mention of genetic manipulation and reproductive intervention was considered extremely scandalous at a time when his friend Julian Huxley was reprimanded by the Director of BBC Radio, Lord Reith, for merely broaching the subject of "birth control" in a radio broadcast.

Haldane's predictions in *Daedalus* have had a deep impact on later developments in molecular biology, genetic engineering, and

reproductive medicine. In 1932, his friend Aldous Huxley incorporated Haldane's ideas in his fictional work *Brave New World*, which
made them known to millions of readers in several generations (see
Dronamraju 1995).

Mathematical Theory of Natural Selection   From 1924 to
1932, Haldane published a series of papers on the mathematical
theory of natural selection (evolution), which has been regarded by
posterity as his most important contribution to science. Haldane's
work (together with the independent mathematical contributions of
R.A. Fisher and Sewall Wright) laid down the foundations of what
later came to be called "population genetics." It also laid down the
theoretical basis for what was later called the "modern evolutionary
synthesis," or simply the "modern synthesis," a term invented by Julian
Huxley when he wrote the book *Evolution: The Modern Synthesis*
(1942). It should be emphasized, however, that although their names
are often mentioned together as founders of population genetics,
Haldane, Fisher and Wright worked independently and pursued different lines of research, although their conclusions were in general
agreement. Haldane, for instance, was a puzzle-solver rather than a
system-builder. On the other hand, Fisher and Wright attempted to
develop systems of their own, resulting in the "fundamental theorem
of natural selection" of Fisher (1930) and the "shifting balance theory"
of Wright (1931). Haldane solved problems as they arose and paid
special attention to human genetics. He was also the most open-
minded of the three founders. As he had no system to defend, he was
always willing to consider other points of view, and was also consistently willing to believe that he might be mistaken.

   Theoretical population genetics laid down the foundations of new
disciplines in science and improved the methodology by introducing
a new rigor, as in the case of epidemiology and physical anthropology.
It provided new tools and knowledge that have greatly contributed to
our understanding of agricultural and medical sciences, such as the
measurement of selection intensity and the impact of mutation on a

population. Haldane's concept of the role of infectious disease as an effective agent in human evolution stimulated a great deal of epidemiologic research, especially for infectious diseases such as malaria.

Other aspects of research, as in quantitative genetics, are based on the underlying concept of treating genes as discrete functioning units of inheritance. These concepts and methods have played an important part in livestock breeding and for improving agricultural crops. In the simplest version, Haldane's early papers assumed a direct relationship between a phenotype and its corresponding gene.

In response to the criticism of Mayr (1959, 1963) and Waddington (1957), who questioned the significance of theoretical population genetics, Haldane (1964) wrote, in his spirited "defense," that his mathematical theory provided the "scaffolding" within which various hypotheses utilizing field data on various species have been tested.

In his mathematical series, Haldane examined the fate of individual genes in populations that may be subjected to varying degrees of selection pressures, different mutation rates, genetic recombination, and ecological and demographic variables, under different types of mating systems such as inbreeding and outcrossing. The impact of other variables, such as the size of the population, as well as the recessivity, dominance, or incomplete dominance, the penetrance of the genes, and the conflict between inbreeding and selection were also considered. These investigations were summed up in his book *The Causes of Evolution* (1932a). In formulating his theory, Haldane (1924) made certain simplifying assumptions:

Although the initial phase of Haldane's work in population genetics was summed up in his *The Causes of Evolution* (1932a), he continued to publish papers in theoretical population genetics throughout his life, some even appearing after his death in 1964.

In his introduction to "A Mathematical Theory of Natural and Artificial Selection," Haldane (1924) wrote:

A satisfactory theory of natural selection must be quantitative. In order to establish the view that natural selection is capable of accounting for the known facts of evolution we must show not only that it can cause a species to change, but that it can cause it to change at a rate which will account for present and past transmutations. (p. 19)

Haldane's early work is discussed in chapter 1.

Gene Fixation   In part V of his mathematical series, Haldane (1927b) investigated the probability of fixation of mutant genes, using the method of generating function suggested by Fisher (1922a). He showed for the first time that a dominant mutant gene having a small selective advantage k in a large random–mating population has a probability of about $2k$ of ultimately becoming established in the population. The probability of fixation is much more difficult to evaluate if the advantageous mutation is completely recessive, but, with remarkable insight, he estimated it as of the order of k/N, where k is the selective advantage of the recessives and N is the population number. Later investigations have fully confirmed the validity of his results. In the same paper, Haldane also considered the equilibrium between recurrent mutation and selective elimination and the rate by which such equilibrium is reached in case of complete dominance. Motoo Kimura (1957) extended these results to include any level of dominance. The probability of eventual fixation, $u(p)$, was expressed in terms of the initial frequency, p, the selection coefficients, and the effective population number. Kimura (1962) later presented a more general formula for $u(p)$ that includes random fluctuations in selection intensity as well as random drift because of small population number.

Metastable Populations   Haldane's (1931) paper on "metastable populations" also deserves special attention. He starts with the assumption that almost every species is in genetic equilibrium, that is to say, with no very drastic changes occurring rapidly in its composition. He stated at the outset that a necessary condition for equilibrium is that all new genes that arise at all frequently by mutation should be disadvantageous; otherwise, they will spread through the population. He cited an example from purple-eyed *Drosophila melanogaster*, where arc wing or axillary speck (each due to a recessive gene) shortened life, but the two together lengthened it.

Haldane considered a population where mutant genes, which are harmful singly, may become advantageous in combination. He then showed that, for *m* genes, any population can be represented by a point in *m*-dimensional space. In many cases, related species represent stable types. The process of speciation can result from a rupture of the metastable equilibrium, and such ruptures may be more likely to occur in small isolated communities. This is very similar to the shifting balance theory that was proposed independently by Wright (1931).

Time of Action of Genes   In a most interesting paper, Haldane (1932c) considered the evolutionary implications of the time in the life cycle of an organism at which genes may act. In general, the more limited the period of action of a gene, the more unalloyed its benefits if it is useful at a certain period. The greater the difference in cell chemistry at different stages of the life cycle, the greater should be the possibility of limiting gene action. Therefore, Haldane argued that, within limits, natural selection will tend to make life cycles more and more variegated, as every increase in complexity will increase the possibility of fixing the time of action of genes. To Haldane, it offered a partial explanation of some of the apparently useless complexities of biology, such as the tendency of parasites to live in widely different hosts.

Another way of fixing the time of action of genes, which was suggested by Haldane,(1932c) is by elimination of part of the chromatin in somatic cells, involving the localization together of all genes that act only in the early stage of development but become redundant later on. Haldane predicted that such a mechanism is what one would "obviously" adopt in designing a "synthetic animal."

Haldane (1932c) concluded that change in the time of action of genes has been an important factor in evolution, and it may explain some cases of orthogenesis, including degeneration. In organisms undergoing metamorphosis, the adaptive efficiency of a gene depends on the limitation in time of its action.

Human Mutation Rates   Haldane was the first to estimate human mutation rates and to show that the mutation rate for certain diseases may be higher in males than in females. While studying the effect of natural selection against the gene for hemophilia on the X chromosome, Haldane pointed out that loss of this gene in each generation must be balanced by recurrent mutation. If this were not so, the disease would quickly die out. Natural selection would act not on normal female heterozygous carriers (XX) but on affected hemizygous males (XY). Thus, Haldane argued that nearly one-third of the known cases of hemophilia must arise by mutation of the gene in each generation.

In *The Causes of Evolution*, Haldane (1932a) estimated that the frequency of mutation for hemophilia is of the order of once in 100,000 generations; that is, p is about 10^{-5} or somewhat higher. Later, he derived the general formula for sex-linked genes as $2\mu + v = (1 - f)x$, where μ and v are the mutation rates in females and males, respectively; x is the frequency of hemophilia among males at birth; and f is the fitness of hemophilic males as a fraction of normal (Haldane 1935). Furthermore, Haldane (1947a) calculated that the mutation rate for hemophilia was much higher in males than in females and made similar observations for sex-linked muscular dystrophy. Haldane's first estimate of mutation rate for hemophilia was 2×10^{-5} (1935) but he raised it to 3.2×10^{-5} in 1947.

In his paper on the impact of recurrent mutation in evolution, Haldane (1933) concluded that mutation is a necessary but not sufficient cause of evolution. Recurrent mutations not only provide the material for selection to act upon but also may give rise to primary and secondary effects, the former due to the accumulation of mutant genes, and the latter to the selective value of conditions that protect the organism against lethal genes. Of special interest is Haldane's emphasis that a certain number might be selectively neutral, anticipating Kimura's (1968) neutral theory of evolution which was proposed many years later.

Genetic Loads   Haldane's 1937 paper on the effect of variation on fitness provided the foundation for the later theory of "genetic loads."

He showed that the effect of mutation on the fitness of a population is independent of how deleterious the mutant phenotype is but is instead determined almost entirely by the mutation rate. The reduction in fitness is equal to the total mutation rate per gamete multiplied by a factor that is between 1 and 2, depending on the dominance of the mutant genes. Haldane's paper provided the first basis for assessing the impact of mutation on the population. It also showed that any increase in mutation rate would have an effect on fitness ultimately equal to this increase. This principle provided a basis for various assessments of the genetic effects of radiation at a time when the question first became socially and politically important. Haldane's principle was independently discovered several years later by Muller (1950), who was deeply concerned with the genetic effects of radiation.

Crow (1958) invented the term "mutation load" for the proportion by which mutation lowers the fitness (or other trait of interest) in an equilibrium population compared to a hypothetical population without mutation. The "segregation load" is the effect of Mendelian segregation compared with a nonsegregating equilibrium population (i.e., one that is asexual). The theory of genetic loads was greatly expanded and discussed by Crow and Kimura (1965) and Kimura and Maruyama (1966).

Quantitative Measurement of Rates of Evolution   Utilizing the quantitative data on tooth measurements in fossil horses, which were published by the paleontologist George Gaylord Simpson (1944), Haldane (1949c) estimated the evolutionary rate per generation. He explained that the dimensions of a solid organ, such as a tooth, a shell, or a bone, are often measurable with great precision. When a series of fossil populations form a lineage, we can calculate the rate of change of the mean value of any measure. To specify a rate, we must have scales for both the time and the character. In evolutionary biology, time is often measured in generations. In the case of a metrical character, the simplest and most accurate measurements are linear. Haldane utilized the fossil data on paracone height and ectoloph length in five fossil horse species.

Haldane (1949c) suggested the term "darwin" for measuring evolutionary rates. This topic was discussed in chapter 2.

Cost of Natural Selection   One of his most brilliant and influential papers in population genetics was published during the last years of his life. It was Haldane's (1957) analysis of the "cost of natural selection" that formed the main justification for Kimura's (1968) "neutral theory of evolution." Haldane investigated mathematically the cost paid by a species per generation during the course of adapting, through natural selection, to changed circumstances such as a change in the environment, an alteration in climate, a new predator, a new source of food, or migration to a new habitat. The cost to the species is expressed in differential mortality and lowered fertility.

Referring to a particular locus, Haldane showed that the total number of selective deaths (or the equivalent in lowered fertility) depends mainly on the initial frequency of the gene that subsequently is favored by natural selection. Haldane showed that the cost incurred by the species, during the process of gene substitution, could be as high as thirty times the population number in a single generation. The number 300 generations is a conservative estimate for a slowly evolving species not at the brink of extinction by Haldane's calculation. For a difference of at least 1,000 genes, 300,000 generations might be needed—maybe more.

Disease and Selection   Haldane's suggestion (1949a, 1949b) that certain genetic polymorphisms might confer immunity to infectious disease stimulated a great deal of research in the epidemiology, genetics, and vaccines for infectious diseases, especially malaria (Dronamraju, 1992, 2004, 2006). Haldane noted the co-occurrence of thalassemia and malaria in countries of the Mediterranean region. He argued that the corpuscles of the anemic heterozygotes may well be more resistant to attacks by the sporozoa that cause malaria than those of either homozygote. His comment was in response to a statement by Neel and Valentine (1947), who suggested that the thalassemia heterozygote is less fit than normal and that the mutation rate may well exceed 4×10^{-4}. Haldane (1949a) noted that if the heterozygote had an increased fitness of only 2 percent, this would account for the incidence without invoking any mutation at all. In support of his hypothesis, Haldane cited examples of similar situations from other species, including *Drosophila*.

Haldane's "malaria hypothesis" was examined and reviewed by Weatherall (2004). Haldane's hypothesis was first confirmed for sickle-cell anemia rather than for thalassemia. It is now known that sickle-cell carriers enjoy almost 80 percent protection against the severe complications of malaria, especially cerebral malaria (Hill et al. 1991). Research in West Africa has confirmed that the relatively high frequencies of HbC (Hemoglobin C) have been maintained by resistance to the malarial parasite *P. falciparum*. The impact of Haldane's hypothesis is summarized in Dronamraju (2006).

Haldane (1949b) further emphasized that certain types of parasitism will tend to encourage speciation and others encourage polymorphism. This is more likely when the parasite is very highly adapted to its host. In certain circumstances, parasitism will be a factor promoting polymorphism and the formation of new species. This will be a random process. Any sufficiently large difference in the times of emergence or oviposition of two similar insect species will make it very difficult for the same parasitoid to attack both of them efficiently. This may give rise to some of the apparently nonadaptive differences between related species.

Haldane's contributions to evolutionary biology, although very important, were only a part of his total scientific output during his lifetime. He devoted many years of his life to teaching and research in physiology, biochemistry and statistics. In addition to the publication of over 400 scientific papers and 24 books, Haldane found time to contribute several hundred popular scientific articles to the lay press throughout his life. This incredible record would be more than enough for any individual, but not for Haldane! He participated in politics, wrote a storybook for children (MyFriend, Mr. Leakey 1937), contributed comic verse regularly each year from 1925 to 1933 to "Brighter Biochemistry", of the Biochemistry Institute at Cambridge University, published a long poem on cancer shortly before his death, and lectured widely in several countries.

6

Ernst Mayr and Evolutionary Biology

Ernst Walter Mayr (1904–2005) was variously described as a zoologist, naturalist, a systematist, and an ornithologist. He was in part all these. They formed the foundation for his career in evolutionary biology. His classic text *Systematics and the Origin of Species* (1942) is one of the pillars of the synthetic theory of evolution. Mayr was a member of the small circle who founded the "modern synthesis," a term invented by Julian Huxley (1942), grandson of the famous Thomas Henry Huxley, who earned the nickname "Darwin's bulldog" for his staunch defense of Darwinism. Julian Huxley and Mayr enjoyed a lifelong friendship. Mayr was a leading founder of the Society for the Study of Evolution in the United States and was the first editor of its journal, *Evolution*.

Beginnings

Ernst Mayr was born on July 5, 1904, in Kempten in Bavaria. His father, Otto, was a jurist in the Bavarian courts but was keenly interested in nature. He used to take Ernst and his two brothers on long field excursions to study natural history. By the time he was ten years old, Ernst was already familiar with the local birds by sight as well as song.

Mayr's long and productive career auspiciously began with a series of fortuitous coincidences in his early life that helped his start in evolutionary biology. These can be phrased in a series of questions beginning with "What if...?" The first of these lucky coincidences occurred when Ernst was only nineteen years old. As a reward for passing the high school examination in February 1923, his mother presented him a pair of binoculars. For several weeks, he cycled daily to the hills outside Dresden, to the hunting chateau of the Saxon kings near the lakes of Moritzburg and to the banks of the Elbe River, indulging in his favorite passion, bird-watching. On March 23, 1923, the first major historical accident in his life occurred. With the aid of his new binoculars, he spotted a pair of ducks in one of the lakes; the male, with a red bill, that was unknown to him. Returning home, he discovered in one of his bird books that what he had observed was a red-crested pochard (*Netta rufina*). What was remarkable about his observation was that the bird had not been seen in Saxony since 1845! There was much discussion among the members of the Saxony Bird Club about the reliability and authenticity of Ernst's observation. One of its members advised Mayr to visit Erwin Stresemann, the leading ornithologist in Berlin. Mayr's career in ornithology, biogeography, and later, evolutionary biology was launched with the encouragement of Stresemann, who was impressed with the young man's enthusiasm for ornithology. Otherwise, he might have completed his medical education at Greifswald to become a doctor, as originally planned.

A number of questions arise in this context. What if Ernst's mother never gave him a binoculars? What if Ernst never spotted those ducks? That was remarkable rare good fortune to observe red-crested pochards that have not been spotted since 1845. What if Ernst never went to Berlin and met with Stresemann? If any one of these possibilities occurred, Ernst Mayr would not have become the famous evolutionary biologist whom we have come to know and admired.

Encouraged by a fellow member of the Saxony Bird Club Mayr traveled to Berlin to meet with the noted ornithologist Erwin Stresemann. After a tough interrogation, Stresemann accepted Mayr's explanation and invited him to join the Berlin Museum as a volunteer during his university holidays. With much enthusiasm, Mayr accepted Stresemann's offer immediately, exclaiming later that "it was as if someone had given me a key to paradise. Bock (2005, 2006, p 171)"

Berlin Museum proved to be an excellent training ground for Mayr. He was introduced to a wide range of avian biology. His future career was going to be in zoology, not medicine as was expected of him according to family tradition. Mayr changed from Medicine at Greifswald to zoology at Berlin in March 1925, completing his Ph.D. in only sixteen months at the age of twenty-one. He had rushed to complete his doctorate to qualify for an assistant's position that was going to be available at the Museum on July 1, 1926.

He was then deeply influenced by Stresemann's ideas on species. Stresemann himself was busy writing his seminal work, *Aves*, during the entire period of Ernst's stay in Berlin. Furthermore, Mayr's ideas on speciation were also influenced by a newly published book by Bernhard Rensch on species concepts and speciation.(Rensch 1929). Mayr always stated that the foundations for his later thinking about the species concept, species taxa, and speciation owe much to Rensch's book (Rensch 1929).

However, most of his major ideas on evolution took shape after he moved to the American Museum of Natural History (AMNH) in New York. There he was able to study the vast collections made by the Whitney South Sea Expedition (WSSE) and the Rothschild collection. These were supplemented by his wide reading of the literature at the library of the AMNH and personal discussions with numerous colleagues and visiting scientists from a wide range of disciplines.

Expedition

The chief attraction for Mayr at the Berlin Museum was Stresemann's promise that he would help Mayr to join an ornithological expedition to the tropics. As I have already indicated, a series of coincidences helped Mayr to join the field expeditions to Dutch New Guinea and Solomon Islands. Perhaps the most important coincidence at that time was Mayr's fortuitous meeting with Walter Rothschild in 1927 through the good will of Erwin Stresemann. Of course, one can play the devil's advocate and say, "What if Mayr had never met Walter Rothschild at that crucial moment in his career?" Another remarkable coincidence was that Lord Rothschild was desperate to find a substitute for his curator of ornithology, Ernst Hartert, who was indisposed. The important point we should remember is that Mayr had the foresight, intelligence, and motivation to make use of these coincidences, channeling his career in a successful direction.

Although at first it seemed foolhardy to send the very inexperienced Ernst on a difficult expedition to a distant land, its success was partly due to the active participation of the Natural History Museum in Buitenzorg (Bogor, south of Jakarta, then the Dutch East Indies). But young Ernst was hardly prepared for such an expedition. He had never collected a bird in his life, nor did he have any experience in handling firearms. Until then, his knowledge of tropical birds and lands was derived only from books. However, with the assistance of three well-trained local museum assistants who knew how to collect and prepare birds and insects, as well as how to camp and survive in the jungle, Ernst became a competent collector and explorer. The director of the Buitenzorg Museum in Jakarta, Dr. K. W. Dammermann, and his assistant Dr. H.C. Siebers provided much needed local support.

That knowledge and experience gained in that expedition stood Ernst well in his personal growth as an evolutionary biologist and field explorer. It became a tripartite expedition; the first two parts were in New Guinea, where he collected for the Rothschild Museum (at Tring, England) and the AMNH and later for the Berlin Museum. The third part of Mayr's expedition played an important role in bringing him to the attention of the Ornithology Department of the AMNH, especially Dr. Leonard C. Sanford, a trustee of the AMNH, who secured funds for bringing Mayr to the AMNH in 1931.

As he was completing the expedition work, Mayr received a telegram inviting him to take part in the WSSE of the AMNH.

Stresemann advised Mayr that accepting that invitation could be good for his future career, which turned out to be remarkably prescient! It is hard to imagine any one matching, let alone surpassing, Mayr's energy and productivity, either in quantity or quality.

Biogeography

Mayr's first and last interest in ornithology was biogeography, as indicated by his doctoral thesis (Mayr 1926) on the northern spread of the serin (*Serinus canaria serinus*) in Europe, and his masterful treatment of the historical biogeography of the birds in northern Melanesia (Mayr and Diamond 2001). The latter is perhaps the most complete geographical analysis of a large group of organisms. His *List of New Guinea Birds* (Mayr 1941) deals with the biogeography of those birds, as well as with their systematics and nomenclature. Over the years, he provided the entire original formulation of "island biogeography" (Mayr 1941), which was tested extensively in his analysis of the birds of the northern Melanesian Islands (Mayr and Diamond 2001). In his treatment of the birds of Timor (Mayr 1944), he was one of the few ornithologists—if not the only one—to apply the (then) new ideas for analyzing the biogeography of birds that were advocated by Stresemann (1939).

Using a faunistic rather than the traditional regional method, Mayr discussed several fundamental aspects such as the definition of a fauna and its origin in a given region, the boundaries of a biogeographic region, and the factors that determine the history of a fauna (Mayr 1944). His method of analysis remains the standard approach today for all research in island biogeography.

During his first decade at the AMNH, Mayr concentrated almost entirely on avian systematics and biogeography. The vast collections of birds from Australia, New Guinea, and the South Pacific were at his disposal. Mayr's research provided him a firm foundation for his theoretical work in systematics and evolution. During his time at the AMNH, Mayr completed his synopsis of New Guinea birds, which is still the basic reference work on this avifauna. He also published his important ideas on island biogeography, some of them appearing in papers with such innocuous titles as "The Birds of Timor and Sumba" (1944).

Mayr's student Walter Bock indicated that the most important result of Mayr's expeditions was a negative one (Bock 2004). Mayr failed to find five of six species of birds of paradise that were known only from trade skins. That failure provided Stresemann with the insight that there was something wrong with those species. Upon careful reexamination of the specimens, Stresemann was convinced that each was a hybrid between two well-known species of birds of paradise (Mayr 1981). Needless to say, none of these discoveries would have been made if Mayr did not take up a career in zoology and ornithology. And what a loss it would have been!

There was no other scientist with the unique qualifications and background of Mayr in South Sea expeditions, avian systematics, and island biogeography. But what is interesting is how this unique background prepared Mayr for his future career in evolutionary studies and its impact on the modern synthesis.

American Museum of Natural History

When Mayr first arrived at the AMNH in New York, he was on a year's leave of absence from the Berlin Museum. His research at the AMNH brought him close to some fundamental problems in evolutionary biology, providing an opportunity to study the species concept, speciation, and biogeography, among others. During his first year, in 1931, Mayr was working with the WSSE material, publishing twelve papers by the end of the year, describing twelve new species and sixty-eight new subspecies. But his appointment was to end soon, except for the intervention of another accidental turn of events. Lord Rothschild, who was being blackmailed because of his indiscretion with a "titled lady," was forced to sell his entire ornithological collection to the AMNH in New York in 1932. Mayr, who was expected to return to Berlin soon, was asked to continue his work at the AMNH, covering both the WSSE and the Rothschild collections. Mayr was appointed as the first and only Whitney-Rothschild curator. He then resigned his position at the Berlin Museum.

Mayr's early years in New York were mainly focused on research in avian systematics and biogeography. The large bird collections from Australia, New Guinea, and the South Pacific provided him an excellent opportunity to study avian speciation, biogeography, and the evolutionary forces involved. This knowledge proved to be a great foundation later for his important speculations and theoretical arguments relating to a whole range of questions in evolutionary biology which Mayr addressed in his great work *Systematics and the Origin of Species* (1942).

Furthermore, he was also responsible for integrating the Rothschild collection into the existing bird collection of the AMNH, which included cataloguing this huge collection and later supervising the Sanford Hall of the Biology of Birds. During those years, Mayr also completed his study of New Guinea birds (1941), and published his ideas on island biogeography (1940), long before MacArthur and Wilson (1967).

Mayr's ideas on the "species concept" led to clear thinking on problems related to speciation, especially his definition of the biological species concept. A central concept of that work was Mayr's exposition of the role of external isolating barriers, which prevent gene flow from one species to others. He argued against the effective role of sympatric mechanisms in speciation, emphasizing that geographic barriers provide adequate mechanisms to account for the speciation process.

While at the AMNH, Mayr made several visits to the Cold Spring Harbor Laboratory, where he enjoyed long discussions on genetics with several leading biologists, such as Theodosius Dobzhansky, Max Delbruck, and Bruce Wallace. It was during those long years that Mayr's ideas on the role of "gene interaction" and "genetic cohesion" in evolution took shape.

Evolutionary Biology (Sympatric Speciation)

While he was still at the AMNH, Mayr became increasingly interested in the evolutionary theory and the species concept. The next and perhaps the final important chance event leading to Mayr's career in evolutionary biology occurred at the 1939 AAAS meeting in December 1939. Mayr was invited to speak on his work on geographical variation and speciation in birds in a symposium on speciation organized jointly by the American Society of Naturalists and the Genetics Society of America. Mayr had the good fortune of speaking immediately after Sewall Wright, whose talk was a disaster, driving away much of the audience. The podium with the microphone was far away from the blackboard. Wright spent much of his time writing long equations on the blackboard and mumbling with his back to the audience. Watching Wright's disastrous talk, Mayr grabbed the sides of the podium firmly and decided to stay at the microphone during his entire talk. The result was an excellent talk that was warmly received by the audience.

Professor L.C. Dunn was in the audience and approached Mayr immediately after the lecture. He invited Mayr to deliver some of the Jesup Lectures in March 1941. Later, Dunn invited him once again to expand his lectures into a book. It was published by Columbia University Press in 1942 under the title *Systematics and the Origin of Species*. The publication of that book secured Mayr's position as a leading figure in the evolutionary synthesis.

After the publication of *Systematics and the Origin of Species* (1942), Mayr's interest turned more and more toward the evolutionary theory. Although he continued to work on birds, he spent more time on the study of evolution and the organization of a Society for the Study of Evolution and its journal, *Evolution*. The correspondence in the appendix contains several letters between Mayr and J.B.S. Haldane on founding the journal. Mayr invited Haldane to join the journal's advisory committee.

As he developed a close friendship with the Haldanes in the post-war years, Mayr visited them in London on more than one occasion. At that time, Haldane was Weldon Professor of Biometry at University College London and his wife, Dr. Helen Spurway, was lecturer in zoology at the same institution. Besides the Haldanes, Mayr cultivated a wide circle of friends who were deeply involved in developing the theory of evolution. One of them was Julian Huxley, who was a cofounder of the Modern Synthesis. Others in his circle included his close friend Theodosius Dobzhansky, as well as Bernhard Rensch, George Gaylord Simpson, Ronald Fisher, and G. Ledyard Stebbins, among others. Another friend was the British ornithologist David Lack (1947) at Oxford.

Mayr's Views on Evolution

As can be expected, Mayr's views on evolution were deeply influenced by Darwin's ideas and writings. In his book *What Evolution Is*, Mayr (2001) wrote that Darwin made a radical break with the typological tradition of essentialism by initiating an entirely new way of thinking by pointing out that the world of living organisms consists of variable populations, not constant classes or types. He wrote:

> Every species consists of numerous local populations. Within a population, in contrast to a class, every individual is uniquely different from every other individual. This is true even for the human species with its six billion individuals. Darwin's new way of thinking, being based on the study of populations, is now referred to as *population thinking*.
>
> This approach was congenial to most naturalists, who in their systematic studies had discovered that species of animals and plants showed as much (and sometimes far more) variation and uniqueness as the human species.... Population thinking is one of the most important concepts in biology: It is the foundation of modern evolutionary theory and one of the basic constituents of the philosophy of biology. (p. 75)

Mayr further wrote: "What is the lowest level of living organization to evolve? ... It is the population. And the population turns out to be the most important site of evolution." (p.76) He explained further: "Evolution is best understood as the genetic turnover of the individuals of every population from generation to generation."(p.76)

Variation and Population Thinking

The ultimate ways of thinking of the population typologist and of the population thinker are precisely the opposite. For the typologist, the type is real and the variation an illusion, while for the populationist the type is an abstraction and only the variation is real. "No two ways of looking at nature could be more different" (Mayr 1959, p. 2).

The availability of variation is a necessary prerequisite for evolution. It is the raw material upon which natural selection can act. Variation that is readily apparent is morphological, such as differences in coat color in mammals, beak size, and shape in birds and wing patterns in Lepidoptera. However, variation can also be seen in physiological traits, biochemical composition, behavior patterns, ecological adaptation, molecular variation, and reproductive performance, to name a few.

In his discussion of speciational evolution, Mayr (1992) wrote that only recently have we understood how different are the concepts to which the term "evolution" has been attached. With wisdom of hindsight, we can now (250 years after Buffon) distinguish three very different concepts of evolution: saltational evolution, transformational evolution, and variational evolution.

Theories postulating saltational evolution are a necessary consequence of essentialism. If one believes in constant types, only the sudden production of a new type can lead to evolutionary change. That such saltations can occur and, indeed, that their occurrence is a necessity are old beliefs. The Darwinian revolution did not end this tradition, which continued to flourish in the writings of Thomas H. Huxley,

William Bateson, Hugo De Vries, J.C. Willis, Richard Goldschmidt, and Otto Schindewolf. Traces of this idea can even be found in the writings of some of the punctuationists.

According to the concept of transformational evolution, first clearly articulated by Jean-Baptiste Lamarck, evolution consists of the gradual transformation of organisms from one condition of existence to another. Almost invariably, transformation theories assume a progression from "lower to higher" and reflect a belief in cosmic teleology resulting in an inevitable steady movement toward an ultimate goal, an ultimate perfection.

Variational Evolution

Mayr emphasized that Darwin introduced an entirely new concept of evolution: variational evolution. New gene pools are generated in every generation, and evolution takes place because the successful individuals produced by these gene pools give rise to the next generation. Evolution thus is merely contingent on certain processes articulated by Darwin: variation and selection. No longer is a fixed object transformed, as in transformational evolution, but an entirely new start is, so to speak, made in every generation. Evolution is no longer necessarily progressive; it no longer strives toward perfection or any other goal. It is opportunistic, hence unpredictable.

What Darwin did not fully realize is that variational evolution takes place at two hierarchical levels, the level of the deme (population) and the level of species. Variational evolution at the level of the deme is what the geneticist deals with. It is effected by individual selection and leads minimally to the maintenance of fitness of the population through stabilizing selection.

The second level of variational evolution is that of the species. Owing to continuing (mostly peripatric) speciation, there is a steady, highly opportunistic production of new species. Most of them are doomed to rapid extinction, but a few may make evolutionary inventions, such as physiological, ecological, or behavioral innovations that give these species improved competitive potential. In that case they may become the starting point of successful new phyletic lineages and adaptive radiations. Such success is nearly always accompanied by the extinction of some competitor. This process of succession of species is often referred to by the term "species selection," but to prevent misunderstandings it may be better to call it "species turnover."

According to Mayr, the transfer from transformational to variational evolution required a conceptual shift that was only imperfectly carried through by most Darwinians. As a consequence, geneticists described evolution simply as a change in gene frequencies in populations, totally ignoring the fact that evolution consists of the two simultaneous but quite separate phenomena of adaptation and diversification. Mayr wrote that the latter results from a process of multiplication of species, a process almost totally ignored in the writings of Fisher, Haldane, Wright, and other leading evolutionary geneticists. This early work in population genetics has been termed "beanbag genetics."

According to Mayr (2001), the geneticists, with the exception of a few saltationists such as Hugo DeVries and William Bateson, usually ignored the problem of speciation altogether. The only geneticists who showed an interest in the multiplication of species were those who had been educated as taxonomists, such as Dobzhansky and Stebbins. The problem of relating speciation to macroevolution occupied primarily three zoologists, Julian Huxley (1942), Mayr (1942, 1954), and Rensch (1959), who were neither geneticists nor paleontologists. Since these three were among the architects of the *evolutionary synthesis*, one can state that the problem of the relation between speciation and macroevolution was not entirely ignored by the evolutionary synthesis (also see Mayr 1970, 1980).

Cost of Evolution

Just as Mayr pointed out that population geneticists have ignored the problem of speciation, there are other problems of evolutionary biology that have been ignored by taxonomists. One such problem has been considered by Haldane (1957b). The process analyzed by Haldane involves a dynamic situation in evolution whenever natural selection favors the replacement of an existing gene by a new allele. Evolutionary change of this kind imposes a substitutional load, and unlike stabilizing selection, this process is associated with the dynamics of evolution. Haldane (1957b) was the first to recognize this process and called it the "cost of natural selection" (discussed further on p. 41). The cost of replacing one allele by another involves a number of deaths that is independent of the intensity of selection. The cost often is so high that evolutionary change is normally an extremely slow process. Mayr suggested that each gene substitution should not be treated as if they were independent of each other. They may, in fact, be partially synergistic when coming together in a genotype. Haldane was, of course, very much aware of this, however. he made certain simplifying assumptions for the sake of mathematical convenience.

Another concern expressed by Mayr was regarding the constancy of selection coefficients assumed by Haldane and Motoo Kimura. However, selection coefficients do change their values from generation to generation, depending on other mortality factors. The overall fitness of a population over long periods may be essentially constant. However, the impact of gene substitution may be mitigated by the fact that the addition of superior new genes or the elimination of deleterious genes could simultaneously change the selective values of the remaining genes, eventually permitting an increase in the rate of substitution. In the case of density-dependent selection, the higher a percentage is killed by gene substitution, the lighter will be the "ecological load" on the remainder of the population. According to Mayr (1963, p. 260), this feedback mechanism guarantees the survival of the population, despite the increased substitution load.

Haldane made some assumptions in an idealized situation, such as that evolution occurs in large populations, that selection intensities are low, and that the initial frequency is determined by mutations. Under these circumstances, one arrives at extremely slow evolutionary rates. However, he was aware that in many natural situations none of these assumptions may be correct. Moreover, the input of new genes into natural populations may be due to immigration, and a new founder population may start off with a relatively high initial frequency of a gene that was rare in the parental population.

"Genetic Cohesion"

Mayr (1959) continued to advance his view that genes do not act alone but are part of a "pool" that exhibits a certain "cohesion" and "harmony." He wrote:

> What gives the local gene pool its cohesion and coadaptation? The work of Dobzhansky (1937) and of others has revealed that there is a harmony among the genes which together make up the local gene pool.
>
> Through natural selection such genes are accumulated and produce maximal viability in their allelic and epistatic interactions. This very general statement is about all we can say about the "coadaptation" or "internal balance" of the gene pool of the local population.

But Mayr (1959) admitted at once that it is not known how such a cohesion is achieved. And curiously, while using such fuzzy terminology as "cohesion" and "harmony," Mayr himself was unable to offer any satisfactory explanation that would provide a more adequate solution to our understanding of the evolutionary theory, that is to say, more adequate than the so-called "beanbag" approach to genetics that he attacked so vehemently. Haldane's reaction to Mayr's argument is discussed in other chapters.

How are such internal balance and coadaptation maintained? Mayr hypothesized that it could be maintained by the multiplicity of interactions of its components, which would automatically resist any disturbance. It would be similar to the state of "genetic homeostasis" that was earlier hypothesized by Lerner (1954); however, also see Schmalhausen (1949).

Genetic Assimilation   Mayr was not the only major scientist who questioned the value of the pioneering work of Haldane, Fisher, and Wright. Conrad Waddington (1905–1975) was an embryologist and evolutionary theorist whose ideas became increasingly popular among the Evo-Devo.

Evolutionary developmental biology (or informally, **evo-devo**) is a field of *biology* that compares the *developmental processes* of different *animals* and *plants* in an attempt to determine the ancestral relationship between *organisms* and how developmental processes *evolved*. It addresses the origin and evolution of *embryonic development*; the role of development and developmental processes in the production of novel features, the role of *developmental plasticity* in evolution; and how *ecology* impacts in development and evolutionary change; and the evolutionary significance of *homology*.

Although interest in the relationship between *ontogeny* and *phylogeny* extends back to the nineteenth century, the contemporary field of evo-devo has gained impetus from the discovery of *genes* regulating *embryonic* development in *model organisms*. General *hypotheses* remain hard to test because organisms differ so much in *shape and form*. Nevertheless, it now appears that just as evolution tends to create new genes from parts of old genes (molecular economy), evo-devo demonstrates that evolution alters developmental processes (genes and gene networks) to create new and novel structures from the old gene networks (such as bone structures of the jaw deviating to the ossicles of the middle ear) or will conserve (molecular economy) a similar program in a host of organisms such as eye development genes in molluscs, insects, and vertebrates. Initially the major interest has been in the evidence of homology in the cellular and molecular mechanisms that regulate body plan and organ development. However more modern approaches include developmental changes associated with speciation.

Community   Like Mayr, Waddington argued against the standard paradigm of population genetics, which focused on frequency changes in genotypes to the exclusion of phenotypic development.

He identified a mechanism, called "genetic assimilation," by which apparently acquired characteristics could appear to be inherited. This would happen when a genetic predisposition to an elevated response to a particular environmental cue during ontogenetic development would be repeatedly selected *for*, until the trait began to appear in the majority of the population independent of the environmental cue. Waddington was also very interested in the role of evolution in shaping human social behavior, both at the genetic level and in the cultural transmission and evolution of learned behaviors, something he termed the "human sociogenetic system."

History of Biology

During the 1970s and later, Mayr's thoughts turned more and more toward the history of biology. This was not an entirely new development as his earlier papers, published during the 1930s, contained historical discussions of the evolutionary theory. He read Darwin's *On the Origin of Species* several times and was quite familiar with Darwin's other works as well. Starting about 1970, he spent several years gathering material for a book on the history of biology but because of the vastness of the subject eventually restricted it to evolutionary biology. It was published under the title *The Growth of Biological Thought* in 1982, a massive volume of 980 pages. It covered in detail systematics, genetics, and evolution.

In his analysis of Darwin's *Origin*, Mayr (1985) pointed out that Darwin had in fact presented an aggregate of five independent theories. One of these theories, which implied that all species or populations of organisms have descended with modification from common ancestors, is historical–narrative. The other four are deductive, and they are evolution as such, gradualism, multiplication of species, and natural selection. Mayr presented a detailed analysis of these theories in his book *Toward a New Philosophy of* Biology (1988). Bock (2004, 2006) presented a summary of Mayr's analysis.

Philosophy of Biology

During the last two decades of his life, Mayr spent much of his time in thinking and writing about the philosophy of biology, especially how it differed from the philosophy of physics. His main point was that all explanations in biology depend on a dual causation, functional and evolutionary. In Mayr's personal view, philosophy of science implied a concept that is restricted to evolutionary biology and perhaps historical evolutionary theory (Bock 2006). Consequently, he believed that biology is made up of concepts rather than laws which are more characteristic of physics. Furthermore, Mayr readily accepted "holism" in biological explanations, while totally rejecting "reductionism." Bock (2006) commented that Mayr's holism was not complete because it did not go beyond the level of the organism, ignoring a completely holistic view of the interaction between an organism and its environment.

Mayr rejected teleology in explaining evolutionary change, accepting accidental cause for phyletic evolutionary change, and it is difficult or impossible to label evolutionary change as goal directed or progressive. Toward the end of his life, Mayr was disappointed that he did not have sufficient time to develop a coherent theory of philosophy of science, especially one that included all aspects of biology. His explanation was that most philosophers of science who were writing on biology lacked an understanding of biology because they were trained in physics or mathematics.

Success as a Biologist

Mayr (1955) himself listed four factors which contributed to his success as a biologist:

(1) His early training under Erwin Stresemann, who taught Mayr how to apply the principles of the new systematics as developed by European taxonomists

(2) His early familiarity with the patterns of geographic variation and speciation of animals inhabiting island regions, resulting from his expedition to New Guinea and Solomon Islands

(3) His work with the extensive collections of the AMNH*, both the Rothschild collection and the Whitney South Seas collection

(4) His rich cultural and intellectual background, which combined both the European scientific tradition and the English-speaking world of the United States

To these we can add other contributing factors:

(5) Mayr's intellectual growth was facilitated by his appointments at two major institutions in the intellectually rich northeastern United States, AMNH★ and Harvard, with their rich scientific resources.

(6) His friendship with the leading evolutionary biologists of that period, which included J.B.S. Haldane, Julian Huxley, and Theodosius Dobzhansky, proved to be immensely stimulating.

(7) His annual visits to the Cold Spring Harbor Laboratory near New York City gave him the opportunity to exchange ideas with numerous eminent scientists, especially in genetics, such as Max Delbruck, Jacques Monod, Boris Ephrussi, Ernst Caspari, and Bruce Wallace. He found them particularly helpful as he had no training in genetics himself.

(8) His success was partly due to his ability to synthesize various concepts, gathering data from multiple disciplines, including ornithology, systematics, evolution, genetics, cytogenetics, geology, geography, paleontology and others. I call this process "intellectual hybridization.", which I discussed in detail in my books,

*American Museum of Natural History

"Haldane", a biographical memoir of J.B.S. Haldane (Dronamraju 1985), and "Foundations of Human Genetics", with special reference to Thomas Kuhn's "paradigm" concept. (Dronamraju 1989)

(9) Although his mother tongue was German, Mayr possessed a great capacity not only to master the English language for his lectures but also to write numerous books in an extremely lucid style that is rare even among native English speakers.*

(10) Finally, his possession of rare good luck to be in the right place at the right time contributed to his success. His sighting of the red-crested pochards that appeared at the right moment, his contact with Stresemann that led to the expedition to New Guinea, and his presence at the AMNH just at the right moment when the museum acquired the extensive bird collection of Walter Rothschild are some examples.

---

*Among the foreign-born scientists who displayed great skill in English language writing, I also include Theodosius Dobzhansky and Guido Pontecorvo, who were born and brought up in Russia and Italy, respectively.

7

Evolution: The Modern Synthesis

A leading architect of the "modern synthesis," Theodosius Dobzhansky, wrote: "The manner of action of selection has been dealt with only theoretically, by means of mathematical analysis. The results of this theoretical work (Haldane, Fisher, and Wright) are however invaluable as a guide for any future experimental attack on the problem" (1937, p. 176). This, indeed, is the answer to the question later posed by Ernst Mayr in his address to the Cold Spring Harbor symposium "Genetics and Twentieth-Century Darwinism" in 1959, which he titled "Where Are We?" (1959).

Another architect of the modern synthesis, Julian Huxley wrote: "Haldane... has made a number of valuable theoretical studies on the rate of evolutionary change to be expected with various given degrees of selective advantage for autosomal dominants and recessives and other types of mutations" (p. 56). Huxley[*] (1942) acknowledged that both R.A. Fisher (1932) and J.B.S. Haldane (1932a) have shown the enormous power of selection in effecting evolutionary change.

[*] See biography of Julian Huxley by Dronamraju (1993).

Haldane's mathematical investigations of natural selection and those of Fisher and Sewall Wright founded the first phase of the modern synthesis. Huxley (1942) summed up Haldane's (1924) early investigations. For instance,

> For ordinary natural selection involving a simple dominant with a selective advantage of 1 in 1,000, it will take nearly 5000 generations to increase the proportion of the dominant from 1 to 50 per cent, and nearly 12,000 more to raise it to 99 per cent. For an advantage of 1 in 100 the number of generations must be divided by 10. For the early stages of selection of a single mutation with constant effects, when the gene is still very rare, dominants can spread much more rapidly than pure recessives, unless a certain degree of inbreeding occurs. (pp. 56–57)

Yet another architect of the modern synthesis, George Gaylord Simpson, wrote:

> The way in which selection and the other factors and forces of evolution interact within a population has been worked out in a brilliant series of studies by R.A. Fisher, J.B.S. Haldane, Sewall Wright, and others. It is these studies, more than anything else, that have made possible the synthesis of generations of observations and experiments in a wide variety of fields into a coherent and comprehensive modern theory of evolution.... No discussion of evolutionary theory need now be taken seriously if it does not reflect knowledge of these studies and does not take them strongly into account.(1950, pp. 225–226)

The title of this chapter is borrowed from Julian Huxley's famous work *Evolution: The Modern Synthesis* (1942). This is a subject that owes its existence to the development of beanbag genetics. Quite simply, the foundations of the modern synthesis were laid by Fisher, Haldane, and Wright in their seminal works on the mathematical theories of evolution during 1918–1934, which have been collectively named "beanbag genetics" by Mayr (1959).Their contributions, including their concepts, methods, and conclusions, formed the foundation upon which later workers were able to design and carry out their evolutionary investigations.

The title *Evolution: The Modern Synthesis* was invented by Julian Huxley (1942) as the title of his landmark publication. In his preface, Huxley explained: "I ... feel sure that a classification and analysis of evolutionary trends and processes as observed or deduced in nature, and the attempted relation of them to the findings of genetics and systematics, is of first-class importance for any unified biological outlook". Huxley acknowledged his great debt to Haldane's (1932a) *The Causes of Evolution*, the two books overlapping in content but also differing considerably in scope and treatment.

If there is any doubt still lingering about the goal of his book, Huxley offered this final explanation in closing his preface:

> The time is ripe for a rapid advance in our understanding of evolution. Genetics, developmental physiology, ecology, systematics, paleontology, cytology, mathematical analysis, have all provided new facts or new tools of research; the need to-day is for concerted attack and synthesis. If this book contributes to such a synthetic point of view, I shall be well content. (p. 8)

Huxley's chapters are arranged in the following order: (1) The theory of natural selection, (2) The multiformity of evolution, (3) Mendelism and evolution, (4) Genetic systems and evolution, (5) The species problem: geographical speciation, (6) Speciation: ecological and genetic, (7) Speciation, evolution and taxonomy, (8) Adaptation and selection, (9) Evolutionary trends, and (10) Evolutionary progress.

Huxley (1942) was very much concerned with the reaction against Darwinism which began in the 1890s and continued later, well into the twentieth century, which he attributed largely to a disconnect between zoologists of successive generations, among other reasons. He wrote:

> The death of Darwinism has been proclaimed not only from the pulpit, but from the biological laboratory.... The major school still seemed to think that the sole aim of zoology was to elucidate the

> relationship of the larger groups. A related school ... spending one's time comparing one thing with another without ever troubling about what either of them really is. In other words, zoology, becoming morphological, suffered divorce from physiology. (pp. 22–23)

Furthermore, evolutionary studies became more and more "merely case-books of real or supposed adaptations" (p. 23).

Huxley lamented that there was little contact of evolutionary speculation with the concrete facts of "cytology and heredity, or with actual experimentation." Huxley presented an admirable account of the development of ideas and concepts, starting with the early Mendelians who worked with clear-cut differences and large effects and assumed that heredity involved sharp and discontinuous mutations. They assumed further that the continuous variation which is widespread in nature is not heritable. Bateson (1902, 1914) introduced the concept of quasi-continuous variations. However, "he relegated selection and adaptation to an unconsidered background."

Next phase saw the rise of antagonism and bitter rancor between the biometricians led by Karl Pearson and Raphael Weldon on the one hand and the Mendelians on the other, until a reconciliation was effected by Fisher (1918) in his famous paper, "The Correlation between Relatives on the Supposition of Mendelian Inheritance." Huxley commented that it was about this time that Mendelism appeared to contradict the facts of paleontology. There was no cohesive attempt to incorporate the discoveries of newer disciplines into a general theory of evolution. Zoologists who followed Darwinian views were looked down on by those who were leading the growth of new disciplines such as genetics, cytology, and comparative physiology.

Later, gradually various facts regarding Mendelian inheritance, segregation and recombination, and mutation were brought together and their role in the evolutionary process understood. Indeed, that was the major contribution of the mathematical theory of evolution which was developed by Haldane, Fisher, and Wright.

Following the foundation laid by the "beanbag geneticists," synthesis came about in several different disciplines: Cytology and genetics became intimately linked, forming cytogenetics; ecology and systematics provided new data and new strategies for studying the evolutionary problems. Gradually, experimental embryology, paleontology, and other sciences joined these studies. The new generation who worked with the famous biologist T.H. Morgan, which included Herman Muller and Alfred Sturtevant, became strong "selectionists" in their evolutionary views. The new and modified Darwinism is far different from the "Darwinism" of Darwin but it is still Darwinism. Huxley concluded his first chapter, "The theory of natural selection," with these words: "It is with this reborn Darwinism ... that I propose to deal in succeeding chapters of this book". (p. 28)

Synthesis of Many Kinds

What has been called "modern synthesis" is, in fact, an amalgam of several different kinds of synthesis. First, there was the grand old synthesis between Darwinian theory of evolution and Mendelian genetics which was led by Haldane, Fisher, and Wright. There were others: (1) synthesis between the three biological disciplines—genetics, systematics, and paleontology; (2) synthesis between an experimental-reductionist approach (genetics) and an observational holistic approach (naturalists and systematists); (3) synthesis between an intellectual tradition with an emphasis on mathematics, selection, and adaptation and a naturalistic tradition with an emphasis on populations, species, and higher taxa. Dobzhansky (1937) led the synthesis between the two approaches—experimental genetics and naturalist-systematics. The evolutionary synthesis also led to a rejection of several anti-Darwinian concepts: (a) the typological-saltational, (b) the teleological-orthogenetic, and (c) the transformationist-Lamarckian theories.

The situation during the period preceding the synthesis was summed up by Mayr (1988). The period between 1859 and 1935 was largely characterized by a gradual unification of a badly split field, consisting of several camps of specialists feuding with each other. There was no abrupt revolution in the Kuhnian sense (Kuhn 1962) but a gradual accumulation of scientific facts from multiple disciplines. I have previously called this process "intellectual hybridization," which represents the creation or hardening of a discipline by a synthesis of concepts, methods, and facts from multiple disciplines (Dronamraju 1985, 1989).

Mayr (1988, p.526) emphasized that the synthesis did not merely accept the principles of theoretical population genetics or what he called the "beanbag genetics." To understand the achievements of the synthesis fully, one need only to appreciate how different the typological and saltationist evolutionary views of the original Mendelians were. What is, however, debatable is Mayr's (1988, p. 526) assertion that the geneticists had to learn from the naturalists about the importance of population thinking. Population thinking, which characterized Darwin's great work, was continued by Haldane, Fisher, and Wright in their mathematical studies of natural selection. Before their monumental work, several others including Hardy (1908) and Weinberg (1908) followed the tradition of population thinking.

What needs to be emphasized is that the underlying genetics of the synthesis is the so-called beanbag genetics, which, contrary to Mayr's (1959) assertion, has played an extremely valuable role in the successful development of the synthesis. Beanbag genetics, in conjunction with systematics, paleontology, and ecology, among other disciplines, formed the backbone of this development.

Nowhere was the synthesis more evident than the development of a unified view of mutations that permitted a reconciliation between the Mendelians and those who worked on quantitative inheritance. It also helped building a bridge between micro- and macroevolution. It showed that the same genetic mechanism was involved in mutations of all magnitudes that represented variants in a continuously varying spectrum. An incidental consequence of this new development was the end of any belief, if it still persisted in a few quarters, in so-called blending inheritance.

The synthesis strengthened the Darwinian concept that all adaptive evolutionary change is a consequence of natural selection that acts upon available variation. It revived and strengthened the role of natural selection in evolution at a time when interest in the role of selection had clearly declined or totally neglected. Both Fisher's *The Genetical Theory of Natural Selection* (1930) and Haldane's *The Causes of*

Evolution (1932a), as well as Wright's (1931) monumental paper, played a decisive role in turning the tide in favor of the recognition of the important role of natural selection.

Other Leaders

Besides Dobzhansky (1937), other major architects of the synthesis are Huxley (1942), Mayr (1942), Simpson (1944), Rensch (1929), and Stebbins (1950). Ford (1940) contributed to our understanding of polymorphism that Huxley (1942, p. 96) incorporated into his modern synthesis.

George Gaylord Simpson   George Gaylord Simpson was responsible for showing that the modern synthesis was compatible with paleontology in his book *Tempo and Mode in Evolution* which was published in 1944. Simpson's work was crucial because so many paleontologists had disagreed, in some cases vigorously, with the idea that natural selection was the main mechanism of evolution. It showed that the trends of linear progression (in, e.g., the evolution of the horse) that earlier paleontologists had used as support for neo-Lamarckism and orthogenesis did not hold up under careful examination. Instead, the fossil record was consistent with the irregular, branching, and nondirectional pattern predicted by the modern synthesis.

In the 1920s and 1930s Fisher, Haldane, and Wright had independently worked out statistical principles by which an advantageous variation could be carried through a population in time and subsequently change the adapted nature of that population. That is, they demonstrated that natural selection could theoretically work. Simpson reconciled these advances in genetics with the fossil record.

Tempo and Mode in Evolution (1944) advanced paleontology in a number of ways. It proposed many means by which evolution might work and demonstrated that hypothesis does have a role in paleontology. It answered the question of whether the fossil record could be reconciled with the new statistical approach that geneticists had applied to natural selection and laid the groundwork for a union of micro- and macroevolution in a single principle. It also showed that the fossil record can be described and interpreted quantitatively. Broadly, it formed part of the greater synthesis that united all the various biological subdisciplines in a common understanding of evolution.

E.B. Ford   Edmund Brisco Ford was an experimental naturalist who wanted to test evolution in nature. He was a student of Julian Huxley and a leading expert in the field of ecological genetics. His seminal work, *Ecological Genetics* (1975), was widely influential. Ford's work complemented that of Dobzhansky. It was as a result of Ford's work, as well as his own, that Dobzhansky changed the emphasis in the third edition of his famous text from drift to selection. Ford was the first to describe and define genetic polymorphism (Ford 1940) and to predict that human blood group polymorphisms might be maintained in the population by providing some protection against disease (Ford 1975).

G.L. Stebbins   The botanist G. Ledyard Stebbins was another major contributor to the synthesis. His major work, *Variation and Evolution in Plants*, was published in 1950. It extended the synthesis to encompass botany including the important effects of hybridization and polyploidy in some kinds of plants.

Mayr's contribution to the evolutionary synthesis was concerned with speciation, especially its origin, causes, and divergence. Species and speciation occupied Mayr's attention all his life. His view was almost exclusively focused on the role of geographic factors in speciation. Mayr's research background was in avian systematics and island zoogeography, an experience he gained through his early expeditions in the Pacific and subsequent work at the American Museum of Natural History (AMNH) in New York. His early training under the eminent ornithologist Erwin Stresemann in Berlin prepared him well for his extensive work at the AMNH and his role in shaping the modern synthesis (Mayr 1942). He was also deeply influenced by his friend Bernhard Rensch, whose book *The Principle of Polytypic Species and the Problem of Speciation* (1929) was the first manifesto of "the new systematics"—Rensch was the first "new systematist." Rensch's later discussions included geographical variation, superspecies, and speciation, as well as the adaptive nature of subspecific differences in birds. Writing to his mentor and friend, Erwin Stresemann, Mayr wrote: "I am presently busy preparing my book manuscript on *Systematics and the Origin of Species*. One cannot deal with this topic without noticing all the time, how much the solution or at least the clear exposition of these problems owes to our friend Rensch" (letter dated June 6, 1941).

Mayr (1942) cited the publications of Rensch more frequently than those of Stresemann and wrote: "My own work was a continuation of the work of Rensch" (Mayr 1976), and "he probably had more influence on my thinking than anyone else; I greatly admired his 1929 book" (Mayr 1980; see biography of Mayr by Jurgen Haffer 2007). (p. 416)

The Modern Synthesis helped to delineate some major points of the evolutionary theory.

(1) Evolution occurs in small gradual steps with a genetic basis. Major contributors are small genetic changes due to mutations or recombination followed by natural selection.

(2) Even small selection forces can be active agents of change.

(3) The rich reservoir of genetic variation, both apparent and hidden, in natural populations is the primary factor in evolution. Genetic diversity in natural populations is the source of variation in populations.

(4) Extrapolation from micro- to macroevolution is proposed.

(5) All evolutionary changes can be explained in a way consistent with known genetic mechanisms and the observed evidence of naturalists.

(6) Geographical factors play a dominant role in speciation.

Harvard evolutionary biologist Stephen J. Gould (2002, pp. 70–71) stated that *The Modern Synthesis* took place in two stages: (a) the initial phase was marked by the pioneering work of the mathematical school which was founded by Haldane (1924, 1932a), Fisher (1930), and

Wright (1931). Their contributions successfully united the Darwinian and Mendelian perspectives.

Fisher, in his seminal work, *The Genetical Theory of Natural Selection* (1930), showed how, slow and gradualist evolution in large, panmictic populations could validate strict Darwinism under particulate inheritance and disprove saltational alternatives by the inverse correlation of frequency and magnitude in variation. To these general chapters, Fisher appended a long chapter on eugenics.

Gould contrasted the initial pluralism of the mathematical pioneers, especially Haldane and Huxley, with the hardening that took place in the second phase. This second phase, as Gould saw it, was more restrictive and displayed an increasingly firm and exclusive commitment to adaptationist scenarios, and to natural selection as a "virtually exclusive mechanism of change." These scenarios according to Gould, are Dobzhansky's 1937 book *Genetics and the Origin of Species*, Mayr's 1942 book *Systematics and the Origin of Species*, and Simpson's 1944 book *Tempo and Mode in Evolution*.

The so-called hardening, according to Gould, was especially evident in the later editions of the founding documents of the second phase. The reasons for the hardening, according to Gould (2002, p. 543), are a complex mixture of empirical and sociological themes with the additional impetus of a cultural analog to drift and founder effects in small populations. Quite perceptively, Gould wrote that the hardening may have been due to the influence of a relatively few leaders. He wrote: "A reassessment by a few key people, bound in close contact and mutual influence, might trigger a general response." (Gould 2002, p. 543) The three leading exponents of hardening in America—Dobzhansky, Simpson, and Mayr worked together as colleagues in a "New York Mafia" centered at Columbia University and the AMNH. Surprisingly, Gould (2002) also included Wright, whose views, according to Gould, altered from initial stress upon stochastic alternatives to selection to an auxiliary role for drift as one aspect of a more inclusive and basically adaptationist process.

Gould (2002, p. 71) pointed out that these two phases of the synthesis, initial pluralism and later hardening, are reflected in the

statements made by leading scientists at the two contrasting centennial celebrations of 1909 and 1959, which marked Darwin's birth and
the publication of the *Origin of Species*, respectively.

Many years later, Mayr (1988, p. 528) commented that when Wright,
in his later papers, had considerably reduced the evolutionary importance of genetic drift, Dobzhansky and Simpson followed that line of
thought in their later publications. But Mayr did not regard this transformation as a sign of hardening of the synthesis, because he wrote that
neither he nor Rensch (nor any other evolutionary biologist) had
attributed to genetic drift much evolutionary significance in the past.

Beanbag Genetics and Modern Synthesis

The genetic foundations of modern synthesis are based on the concepts and methods of beanbag genetics. Indeed, as Mayr was one of
the architects of the synthesis, he was surely aware of the genetic
foundations of the modern synthesis even though his own contribution to the synthesis was in systematics.

Under the chapter heading "Mendelism and evolution," Huxley
(1942) wrote:

> The essence of Mendelian heredity is that it is particulate. The genetic
> constitution is composed of discrete units. Each kind of unit can exist
> in a number of discrete forms. The hereditary transmission of any one
> kind of unit is more or less independent of that of other units.... The
> units are the Mendelian factors or genes, while their different forms
> are called allelomorphs or alleles. (p. 47)

Much of Huxley's discussion of evolution in modern synthesis was
gene-based and it was "beanbag" genetics of the kind which Haldane
(1964) defended in his famous defense. Huxley wrote:

> Of Mendelizing differences, alike in domestic and wild species, which
> are actually or potentially favourable.... We need only to think of the

genes producing small and large size in poultry..., those producing the specific differences between the two species of snapdragon..., the mimetic polymorphic forms of various butterflies..., the different forms of heterostyled flowers such as primrose, the single-brooded and double-brooded condition in silk worms; and so on. (p. 53)

Huxley described that genes are in many ways as unitary as atoms, and they vary in their action in accordance with their mutual relations. Huxley, like Haldane, Fisher, and Wright, emphasized the independent hereditary behavior of genes, from which their discreteness is deduced. Huxley compared chromosomes with "supermolecules" that are built up out of a series of regions, "each region marked off by zones of potential breakage." (p. 49)

Earlier research by Fisher, Ford, and Huxley (1939) showed that in chimpanzees not only are the same differences found, but tasters and nontasters of PTC (Phenyl Thiocarbamide) occur in about the same proportions as in man—close to 3:1. As Huxley noted, this indicates a stable balance between the two conditions, perhaps involving a heterozygote advantage. Huxley cited various blood-group genes that may also fall into a similar category, occurring in different proportions and in different species of lower mammals.

Huxley's discussion of modern synthesis is typical of what would expect in a "beanbag" type of discussion. "Many differences between 'good' species have also been shown to depend on mendelizing gene-differences." (p. 53) Huxley's discussion was heavily based on the previous work of Fisher (1930) and Haldane (1932a), which was termed "beanbag genetics" by Mayr (1959, 1963). Huxley (1942, p. 56) quoted Haldane's (1932a) "valuable theoretical studies" on the rate of evolutionary change to be expected under various degrees of selective advantage for autosomal dominants, recessives, and other types of mutations. One important conclusion is that intense competition favors variable or plastic response to the environment rather than high average response. This would explain the large variability found in many natural populations.

Genes and Characters

According to Mayr (1959), one of the principal tenets of beanbag genetics is the direct association between a gene and its corresponding phenotype. Huxley (1942) noted that although the notion of Mendelian characters has been entirely dropped, it may still be useful occasionally. His knowledge, of course, reflected the state of genetics at that time. Huxley noted that it is a false conception for several reasons.

In the first place, a single gene may affect several characters (pleiotropism). Second, a gene may exert a direct effect on a single process, but in many different sites and conditions. Third, a gene exerts a primary direct effect, which then causes numerous secondary effects. But nowhere did Huxley expound the concepts that Conrad Waddington and Mayr took up later on. In *Evolution: The Modern Synthesis*, we find no reference to "genetic cohesion," "coadapted harmony of the gene pool," "integrated gene complexes," or "unity of the genotype" that Mayr (1959, 1963) stated as better alternatives to "beanbag" genetics. However, Huxley (1942) referred to other complications such as gene–environment interactions, modifying genes and polygenic characters studied by Mather (1941).

In his earlier work, Dobzhansky (1937), too, did not mention this terminology that later became so important to Mayr. Indeed, the so-called shortcomings of the "mathematical foundations of evolution," which troubled Mayr so much, did not seem to bother Dobzhansky at all. On the contrary, Dobzhansky (1937) wrote of the mathematical contributions of Haldane, Fisher, and Wright approvingly, "The manner of action of selection has been dealt with only theoretically, by means of mathematical analysis. The results of this theoretical work [Haldane, Fisher, Wright] are however invaluable as a guide for any future experimental attack on the problem" (p. 176). In these few words, Dobzhansky (1937) succinctly summed up the significance of the mathematical theory of natural selection and, in doing so, inadvertently provided the answer to Mayr's (1959) challenge that was put forward at the Cold Spring Harbor symposium, several years later.

While going through *Evolution: The Modern Synthesis*, one finds, in contrast to the skepticism of Mayr (1959) and Waddington (1953), the following appraisal of *The Causes of* Evolution by Julian Huxley:

> Haldane ... has made a number of valuable theoretical studies on the rate of evolutionary change to be expected with various given degrees of selective advantage for autosomal dominants and recessives and other types of mutations. One important conclusion is that intense competition favors variable or plastic response to the environment rather than high average response. This presumably helps to explain the large variability to be found in many natural populations. (Haldane 1932a, p. 56)

It is perhaps worthwhile to consider Wright's (1968a) commentary on Haldane's mathematical theory of natural selection for at least two reasons. With Haldane and Fisher, Wright was a cofounder of population genetics, which brought Mendelian genetics and Darwinian evolution together. Second, unlike Dobzhansky, Huxley, and Mayr, Wright was mathematically gifted and thus fully able to evaluate Haldane's contributions. With respect to Haldane's analysis of the effect of natural selection, Wright (1968a) commented that at the time of Haldane's (1924) work, the precise effect of selection on the composition of a Mendelian population had been presented in only the simplest cases.

Haldane proceeded to make a comprehensive investigation of the subject. At first he considered thirteen relatively simple situations for which he gave formulas for the number of generations required to produce any given change under various genetic situations. In later parts, he investigated more complex situations that were not acknowledged by Mayr (1959) in his critique of the founders of mathematical theory of evolution. They include, among others, the effects of self-fertilization and less intense inbreeding, assortative mating, and selective fertilization; the effects of selection on systems of multiple interacting factors, on metastable systems, and on polyploids; the consequences of overlapping instead of discrete generations; the balancing of adverse selection by recurrent mutation; and the effects of partial isolation and rapid selection and of partial isolation and rapid selection.

Elsewhere, Wright (1968a) wrote: "Haldane's knowledge of physiology and Biochemistry enabled him to discuss the relations of genes to characters in a fruitful way in many papers" (p. 3). Wright and Fisher did not possess that wider knowledge.

Modern Synthesis and Target of Selection

Mayr (1963, 1988) was eloquent on the subject of "target" of selection. During Darwin's time and later, the organism as a whole, rather the entire phenotype, was considered the target of selection. It was during the years following the work of Bateson (1906) and Johannsen (1909) that the distinction between phenotype and genotype became a problem. Mayr (1988) made a distinction between the German geneticists who treated both the phenotype and the genotype holistically, with an emphasis on genic interactions. Consequently, the individual was treated as the target of selection as was the case in Darwin's discussions.

In contrast, in the English-language scientific literature, a strong reductionist trend prevailed, which led to the study of genes as independent isolated units. Even before the advent of mathematical population genetics, genes were studied in more or less isolation by members of the Morgan school at Columbia University, especially by Muller and Sturtevant. That tradition was followed in their early mathematical papers by Haldane, Fisher, and Wright. Mayr (1988) commented that those who accepted the change of gene frequencies as the earmark of evolution ignored whether such change was due to genetic drift or to selection. And, he added, "in the single-minded concentration on genes, it was often forgotten how important individuals, populations, and species are in determining the course of evolution". (p. 102)

Citing Sober's (1984) analysis, Mayr (1988, p. 101) argued that the question to be asked is "selection of," not "selection for." And this, explained Mayr (1988), makes it clear why the individual and not the gene must be considered the target of selection. Mayr's reasons are that (a) it is the individual as a whole that either does or does not have reproductive success; (b) the selective value of a particular gene may vary greatly depending on the genotypic background on which it is placed; (c) since different individuals of the same population differ at many loci, it would be very difficult to calculate the contributions to each of these loci to the fitness of a given individual; and (d) accepting the individual as a whole leads to the confusing distinction between internal selection (dealing with processes occurring during ontogeny) and external selection (dealing with the interaction between the adult and the environment).

The discovery of enormous variation in enzyme genes in the 1960s led Motoo Kimura (1968) to propose the theory that most changes in gene frequencies are of no selective significance but are neutral. Although natural selection continued to be considered a factor in the evolutionary change, Kimura and his followers have certainly reduced its importance in that process.

Dobzhansky's Book   Theodosius Dobzhansky, who migrated to the United States in 1927 from the Ukraine, joined Morgan's fruit fly lab at Columbia University as a postdoctoral fellow supported by the Rockefeller Foundation. Working mostly with *Drosophila pseudoobscura*, he pioneered the genetic studies of natural populations.

In his early research in the Soviet Union, Dobzhansky was influenced by the work of the Russian geneticist Sergei Tschetwerikoff (Chetverikov[1]1926) on the role of hidden genetic variability in the evolution of natural populations, and the role of recessive genes in maintaining a reservoir of genetic variability in a population (Dobzhansky 1933).

---

[1] I use the Russian spelling that was used by Dobzhansky (1937). Mayr (1988, 1991) used a different version, Chetverikov, in his books.

Dobzhansky's (1937) seminal work, *Genetics and the Origin of Species*, was a deliberate attempt to bridge the gap between population genetics and natural history. It was described in the preface, written by Leslie Dunn, as signalizing something that can only be called the "back-to-nature" movement. Dobzhansky pioneered experimental investigations in the field, involving mostly *Drosophila* species, studying genetics and evolution, which until then was investigated mainly by means of laboratory experiments or mathematical equations (See Dobzhansky 1973).

Dobzhansky's studies revealed that natural populations had far more genetic variability than was realized at that time. For those who found the early mathematical studies of evolution difficult to understand Dobzhansky's nonmathematical and lucid account of evolutionary genetics was helpful. His 1937 book was the second nonmathematical account of evolutionary genetics, the first being Haldane's *The Causes of Evolution*, which was published five years earlier.

Furthermore, Dobzhansky's ideas in evolutionary biology were influenced by his close friendship with Haldane during the 1930s. Haldane was a visiting professor with Dobzhansky at Caltech during 1932–1933. Many years later, it was Haldane's turn to arrange a visit by Dobzhansky in 1959 to the Indian Statistical Institute in Calcutta, where he was then employed as a research professor.

Dobzhansky (1937, 1951) emphasized the differences as well as the interrelatedness between the disciplines of genetics and evolution: "It should be reiterated that genetics as a discipline is not synonymous with the evolution theory, nor is the evolution theory synonymous with any subdivision of genetics. Nevertheless, ... any evolution theory which disregards the established genetic principles is faulty at its source" (p. 8).

Isolating Mechanisms Dobzhansky (1937) identified different kinds of isolating mechanisms. They may be directed at preventing the production of hybrid zygotes, or to ensure that no hybrid reaches the reproductive stage. There may be ecological mechanisms such as the confinement of potential parents to different habitats so that they are kept apart during the reproductive season, or seasonal or temporal isolation, involving maturation of the reproductive parts at different times of the year.

In other situations, parental forms occur together but hybridization is excluded because of (a) sexual or psychological isolation because of some form of incompatibility between potential mates, (b) mechanical isolation that involves physical incompatibility of the reproductive organs, (c) physiological problems that prevent the union of spermatozoa with the eggs, and (d) inviability of the hybrids— fertilization takes place, but the zygote dies at some stage of development before it becomes a sexually mature organism. Finally, hybrid sterility prevents the reproduction of hybrids that have reached the breeding stage. Sterile hybrids produce either no functional gametes, or gametes that give rise to inviable zygotes.

Population genetics   Later, Dobzhansky (1937) discussed different types of evolution based on the magnitude of change involved, for instance, micro- and macroevolution. It was while considering the nature of changes occurring in the genetic composition of populations he made the following statement, introducing the term "population genetics": "Since evolution is a change in the genetic composition of populations, the mechanisms of evolution constitute problems of population genetics" (pp. 11–12). Dobzhansky introduced the term "population genetics" so casually that is easy to overlook it. There was no indication that it represented an important and rapidly growing branch of genetics. Dobzhansky argued that natural selection worked to maintain genetic diversity as well as driving change.

Dobzhansky's unique ability to combine genetics, cytology, ecology, and natural history attracted many other biologists to join him in his effort to find a unified explanation of how evolutionary processes occur in nature. Foremost among these were Mayr (1942), Huxley (1942), Simpson (1944), and Stebbins (1950). Their combined work brought together genetics, cytology, paleontology, systematics, and many other sciences into a coherent explanation of evolution. The modern synthesis became the foundation for future research. Dobzhansky's (1937) *Genetics and the Origin of Species* captivated several other biologists. Mayr found the book to be an enormous inspiration; he was then deeply involved in the classification of great bird collections at the AMNH and the discovery of new bird species.

Dobzhansky's (1937) book is usually considered the first major work of neo- Darwinism, which appeared in the postmathematical years. Mayr (1982) stated that this publication "heralded the beginning of the synthesis, and in fact was more responsible for it than any other." Francisco J. Ayala (1977), a leading pupil of Dobzhansky, wrote:

> Dobzhansky's most significant contribution to science doubtless was the modern synthesis of evolutionary theory accomplished in *Genetics and the Origin of Species*, first published in 1937—a book that may be considered as the twentieth century counterpart of Darwin's *The Origin of Species* (1859). The title of Dobzhansky's book suggests its theme: the role of genetics in explaining the origin of species, i.e. a synthesis of genetic knowledge and Darwin's theory of evolution by natural selection.

Major synthesis of Darwinian evolution and Mendelian genetics was admirably accomplished by the pioneers, Haldane, Fisher, and Wright, in their mathematical theories of selection, during the years 1924–1934. Of these, Haldane's *The Causes of Evolution* (1932a) was the only nonmathematical account of synthesis at that time. In addition to the genetic and evolutionary considerations, Haldane also incorporated evidence from ecology, cytogenetics, and interspecific crosses.

Dobzhansky's *Genetics and the Origin of Species* took this further. Dobzhansky presented a nonmathematical account, citing evidence from both laboratory experiments as well as data collected in the field. He paid special attention to hybrid sterility and species formation. Both Haldane (1932a) and Dobzhansky (1937) discussed the role of polyploidy, especially allopolyploidy, in evolution.

Earlier, Dobzhansky (1935) proposed the following definition of a species: "That stage of evolutionary process, at which the once actually or potentially interbreeding array of forms becomes segregated in two or more separate arrays which are physiologically incapable of interbreeding." Mayr (1940) offered a slightly different definition of the biological species concept:

> Species consists of a group of populations which replace each other geographically or ecologically and of which the neighboring ones intergrade or hybridize wherever they are in contact or which are potentially capable of doing so (with one or more of the populations) in those cases where contact is prevented by geographical or ecological barriers. (p. 256)

While discussing the definition of species in *Systematics and the Origin of Species* (1942), Mayr offered a shorter version: "Species are groups of actually or potentially interbreeding natural populations, which are reproductively isolated from other such groups".

Beanbag Genetics, Evolution, and Speciation

Mayr's course of lectures at the University of Minnesota in 1949 was titled "Evolution and Speciation." They became the very first draft for a new book that appeared several years later, in 1963, under the title *Animal Species and Evolution*. It was a comprehensive summary of the biogeography and natural history of species and speciation and a grand synthesis of population genetics, the origin of new species, and adaptive radiation.

The major unifying theme of Mayr's book is the impact of population thinking on our understanding of the species problem and the evolutionary process. The historiography of several evolutionary topics is refreshing. Mayr summarized the evidence for allopatric speciation. Integrated nature of the gene pool of a population is emphasized once again. Reproductive isolation of a species protects against the disruption of its well-integrated coadapted gene system.

Among other topics discussed by Mayr are the biological species concept, sibling species, allopatric speciation, isolating mechanisms and their occasional breakdown, and genetic revolutions. Such a revolution is defined by Mayr as a drastic genetic change that may occur in a natural population, affecting all loci at once. Another aspect of evolution discussed by Mayr (1963) is that the process of evolution above the species level is a continuation of the factors responsible for the origin of subspecies and species; that is to say, macroevolution is a continuation of microevolution.

Genetic Revolutions and Beanbag Genetics Mayr's contribution to *Evolution as a Process* (Huxley et al. 1954) is memorable for his interesting idea of "genetic revolutions." In an article with the title "Change of genetic environment and evolution," Mayr (1954) discussed the role of various genetic processes which occur during speciation of peripherally isolated founder populations. He was in particular interested if the limited gene pool in such populations would result in a different set of selective values (of the genes) from that of the parent (founder) population. This is in part due to the increase in inbreeding that is expected in such populations and the resulting homozygosity. Among other factors, Mayr was especially concerned with the impact of gene flow and of isolation.

Mayr referred to classical genetic studies where genetic changes at each locus were studied separately in order to simplify the analysis. In Mayr's words, "The genetic factors of an organism were treated as so many beans in a large bag. That this is not so is now known to every geneticist, but 'bean-bag' thinking is still widespread. The fact is, of course, that genes do not exist in 'splendid isolation,' but are parts of an integrated system" (p. 164).

Mayr drew attention to the fact that the action of a given gene is strongly influenced by its genetic background, its genetic "coworkers." And what is true for the function of a gene is true also for its "selective value." Mayr concluded: "The selective value or viability of a gene is thus not an intrinsic property but is the sum- total of the viabilities on all the genetic backgrounds that occur in a population?" (p. 165)

The fact that the viability of a given allele depends on its genetic background has been discussed earlier by both Wright (1931) and Dobzhansky (1950). Mayr (1954) presented a long argument in favor of a "well-integrated, coadapted gene-complex" that constitutes an evolutionary unit despite its intrinsic variability. Any disharmonious gene or gene combination that attempts to become incorporated in such a gene complex will be discriminated against by selection.

Mayr (1954) hypothesized that during a "genetic revolution" the population will pass from one well-integrated and rather conservative condition through a highly unstable period to another new period of balanced integration. Furthermore, it "may well have the character of a chain reaction. Changes in any locus will in turn affect the selective values at many other loci, until finally the system has reached a new state of equilibrium" (p. 170). In other words, small founder populations may pass through a highly unstable period to another period of balanced integration (Mayr 1963, p. 538). During that process the population will lose much of its genetic variability. Mayr (1954) cited examples of natural populations which pass through a *genetic bottleneck* of reduced variability. One such case is the introduction of the European starling to the United States in 1890. Of less than twenty pairs introduced, only a fraction of them bred successfully. After several years of slow growth, their population expanded greatly, reaching about 50 million individuals by 1954. Other examples mentioned by Mayr include the house sparrow, the Japanese beetle, and the potato beetle.

Mayr recognized that the period of genetic reorganization and relaxed selection pressure is not only a period of rapid evolutionary change but also offers an exceptional opportunity for a drastic ecological change of a partially unbalanced genetic milieu. It is also an occasion when many evolutionary novelties may appear.

According to the biography of Mayr by Haffer (2007, p. 219), Mayr got the idea of "genetic revolution" in peripherally isolated populations while visiting Naples, Italy, during the summer of 1951. Shortly before that visit, Mayr lectured at Oxford when E.B. Ford invited him to contribute a chapter to the book *Evolution As a Process* that was published in 1954, and "genetic revolution" was introduced in that article.

Mayr (1942) versus Mayr (1963)

Gould (2002) questioned Mayr concerning the differences in emphasis between Mayr's 1942 book *Systematics and the Origin of Species* and his 1963 book *Animal Species and Evolution*. The publication of the two books was separated by some twenty-one years, and Gould wondered if Mayr's views on evolution were transformed in some fundamental fashion during that intervening period. In particular, Gould was referring to a greater emphasis on the role of adaptation in the later volume and a similar emphasis on the origin and development of diversity in the 1942 book. In a private letter to Gould, dated December 20, 1991, Mayr wrote:

> I consider evolution by and large to consist of two processes: (1) the maintenance and improvement of adaptedness, and (2) the origin and development of diversity. Since (2) was almost totally ignored by the pre-synthesis geneticists Haldane, Fisher and Wright, I focused in 1942 on (2). By the 1950s the study of diversity had been fully admitted to evolutionary biology, owing to the efforts of Dobzhansky, myself, Rensch and Stebbins, and in my 1963 book I could devote a good deal of attention to (1). This was rather easy because, as you know, I used to be a Lamarckian. And Lamarckians are adaptationists. Hence, it is not that from 1942 to 1963 I had become an adaptationist, rather I reconciled in 1963 my adaptationist inclination with the Darwinian mechanism.

A related question that is often asked is whether Mayr had become a stronger selectionist after 1942. This view was supported by the fact that several cases considered as examples of neutral polymorphism in 1942 were reinterpreted in the 1963 book as selectively balanced polymorphism because of the absence of large fluctuations.

According to Gould (2002), this difference provides a fascinating illustration of how scholars can slowly and unconsciously imbibe a shifting professional consensus, thus imposing a "subjective and personal impression of stability upon a virtual transmogrification" (p. 536). Gould commented further that he found this unconscious alteration all the more ironic because Mayr's first category of major change in ideas about speciation—his intellectual move from the dumbbell to the peripatric model—so strongly encourages a widened space for nonadaptationist themes.

Significant Stages in the Modification of Darwinism

Date	Stage	Modification
1859	Darwinism	Population thinking, evolution
1883, 1886	Weismann's neo-Darwinism	Diploidy and genetic recombination recognized
1900	Mendelism	Particulate inheritance accepted, blending inheritance rejected
1918–1934	Population genetics	Evolution measured in gene frequencies; small selection pressures are considered effective agents of change; genetic drift
1937–1950	Modern synthesis	Multidisciplinary approach to the study of evolution; evolution of diversity
1950–1970	Postsynthesis	Increased recognition of stochastic processes in evolution, punctuated equilibria; altruism; cost of evolution
1969–1980	Rediscovery of sexual selection	Importance of reproductive success for selection; game theory
1965–present	Molecular evolution	Protein polymorphism, neutral theory

(Modified after Mayr 1991)

Postsynthesis Period

From Mayr's (1988) point of view, the following revisions of the Darwinian theory have occurred during the postsynthesis period. Some of these are generally accepted but others have generated and continue to generate debate and disagreement among evolutionary biologists and geneticists:

(1) The individual organism is the principal target of selection.

(2) Genetic variation is largely a chance phenomenon; stochastic processes play a major role in evolution.

(3) The genotypic variation that is exposed to selection is primarily a product of recombination, and only ultimately of mutation.

(4) All speciation is simultaneously a genetic and a populational phenomenon.

(5) "Gradual" evolution means primarily populational evolution, but it may include major phenotypic discontinuities.

(6) Evolution means changes in adaptation and diversity, not merely a change in gene frequencies.

(7) Selection is probabilistic, not deterministic.

(8) Every stage in the life cycle of an individual is exposed to natural selection.

(9) Many genetic changes, primarily at the molecular level, are probably neutral or nearly neutral.

(10) Selection can operate in kin groups through inclusive fitness.

(11) The history of life has been drastically affected by major extinctions that may have fallen largely at random. (Modified after Mayr 1988, p. 532)

Beanbag Genetics Continues

The modern synthesis of evolution continues today, having been refined after the initial establishment by 1950. Beanbag genetics continued in the works of several biologists, including W.D. Hamilton and John Maynard Smith, who continued the development of a gene-centric view of evolution in the 1960s. Later advances in the synthesis as it exists now have extended the scope of the Darwinian idea of natural selection to include subsequent scientific discoveries and concepts unknown to Darwin, such as developments in DNA technology and genetics that allow rigorous experimental and mathematical analyses of phenomena such as kin selection, altruism, and speciation. Other recent advances are discussed on p. 187.

The gene-centric view emphasized that the gene is the only true unit of selection. Dawkins (1976) extended the Darwinian idea to include nonbiological systems exhibiting the same type of selective behavior. The synthesis continues to undergo transformation. Recent years saw another area of intensive research that concerns the relative importance of natural selection and genetic drift, the neutral theory of molecular evolution, and the nearly neutral theory of molecular evolution.

Contrary to the gene-centric view, Mayr argued that the individual as a whole rather than each separate gene is the target of selection. He saw the genotype to be a well-integrated system "analogous to an organism with structure and organs." Mayr (1988) (p. 101) wrote that this view was developed by the Russian geneticists Ivan Schmalhausen and Dobzhansky and their collaborators, as well as Waddington and Rensch. It was forcefully presented in the theory of "genetic homeostasis" that was developed by I.M. Lerner (1954) and in Mayr's (1963, 1976) own writings. Mayr (1988) emphasized once again that the interactions within the genotype appear to be far more complex than was apparent from the work presented in classical population genetics.

Nevertheless, as Crow (2008) had emphasized, beanbag genetics is much more than what Mayr had characterized. Much of the recent work in molecular genetics and developmental genetics has continued that tradition.

8

Summary

J.B.S. Haldane (1964) opened his defense of beanbag genetics with the following sentence, "My friend Professor Ernst Mayr, of Harvard University, in his recent book *Animal Species and Evolution* ..., which I find admirable, though I disagree with quite a lot of it." Thus, Haldane summed up his friendship with Mayr in an appropriate and precise manner—they enjoyed an intimate friendship on the personal level but continued to disagree on the intellectual plane on certain aspects of evolutionary biology. It must be said at the outset that as controversies go, this was a mild one. There has never been any personal rancor between them, unlike the situation between R.A. Fisher and Sewall Wright. Indeed, as the correspondence indicates, Haldane and Mayr displayed friendship and respect for each other. Their differences were mostly concerned with emphasis on certain issues rather than a total disagreement on major issues in evolutionary biology. Both agreed that natural selection was the main driver of change in evolution. Several years later, in a review of the reprint of Haldane's *The Causes of Evolution*, Mayr (1992), to his surprise, found that they were closer in their views on certain issues, for instance, the pitfalls of studying genes in isolation, than he had realized at the time of his address to the Cold Spring Harbor symposium "Genetics and

Twentieth-Century Darwinism" in 1959. Nevertheless, we must be grateful to Mayr (1959) for challenging the great founders of population genetics; otherwise, Haldane would not have written his brilliant essay, "A Defense of Beanbag Genetics," which was published shortly before his death in 1964.

In a letter dated July 31, 2002, addressed to me, Mayr wrote:

> I always tell everybody that Haldane was the most brilliant person I ever met in my life; considering his enormous gifts, it has always been a puzzle to me that he has not made more great discoveries that would be recorded in a history of biology. I think he had too many interests and never concentrated on any specific problem in evol. biology.

Reading through the correspondence, it is clear that Mayr's admiration for Haldane was never in doubt. Haldane, too, enjoyed the exchange with Mayr.

Their personal discussions were cordial and a joy to behold, especially for a young student, as I was then, covering a whole range of scientific subjects. Those personal meetings mostly consisted of Mayr asking Haldane a series of questions and Haldane answering them patiently. Naturally, evolutionary biology and population genetics received special attention. Haldane also enjoyed his meetings with Mayr and offered warm hospitality when Mayr visited him in London and Calcutta. He made special arrangements when Mayr visited India in 1959 which included the Calcutta zoo, bird watching trips as well as visiting other places of scientific and cultural interest in West Bengal and Orissa.

Haldane and Mayr differed in age, background, education, and research interests. But they shared a common interest in evolutionary biology. Mayr was born in 1904; Haldane was born in 1892. By the time Mayr obtained his Ph.D. in zoology in Berlin, Haldane was already a famous scientist and author, having published several papers on the mathematical theory of natural selection and his 1923 classic *Daedalus; or, Science and the Future* in which he predicted future developments in molecular biology.

There are other differences, especially in their background and intellectual outlook, which provide an explanation of sorts in their differing approach to the study of evolution. Haldane's strength was in mathematics. He went to New College (Oxford University) on a mathematical scholarship and knew enough math to provide support for physiological experiments of his father, Oxford University physiologist John Scott Haldane. Starting with his first physiological paper in 1912 and his first genetics paper in 1915, Haldane applied his mathematical facility in his research. Another significant fact is that Haldane's degree from Oxford was in classics. In fact, he never received a formal education in science. His early scientific education was provided by his father, and the rest was self-taught.[*]

Mayr, on the other hand, received a formal education in zoology, receiving a Ph.D. under the direction of the noted ornithologist Erwin Stresemann at the University of Berlin. Mayr's father, too, encouraged his sons to appreciate natural history, taking them on regular Sunday walks in the country. Mayr later wrote: "It was on these trips that we collected flowers, mushrooms, fossils ... or did other natural history studies. When my father heard of a heron colony ... near Munich, we also visited it" (from Haffer 2007, p. 12). As he got older, Mayr's bird-watching skills improved steadily, culminating in his important discovery of red-crested pochards (discussed in detail on p. 131).

While Haldane came to view evolutionary problems from a quantitative and mathematical point of view, Mayr's interest grew out of his interest in ornithology and island biogeography as well as speciation which is derived from his early expeditions to New Guinea and the Solomon Islands. His expedition was crucial in transforming Mayr's life. He started as a novice and ended as a seasoned field man (Bock 2006).

---

[*] Haldane's interest in ecology should be mentioned. This was particularly evident in his later years when he joined Mayr in bird watching in India and enjoyed a close friendship with the Indian Ornithologist Salim Ali (1985).

Mayr's extensive work in the taxonomy of various bird collections from the Pacific at the American Museum of Natural History (AMNH), especially his first "love", the birds of New Guinea, provided a sound basis on which later authors could build. Based on the biological species concept, it provided an overview of the distribution of all the bird taxa of that region, later leading to the consideration of general zoogeographical topics and historical-dynamic analyses of world faunas.

Mayr had no formal training in genetics. Much of his genetic knowledge came to him later, partly due to his contacts with geneticists when he spent several summers at the Cold Spring Harbor Laboratory, where he found a congenial intellectual environment. His, education in genetics was furthered by his close association with Theodosius Dobzhansky, Bruce Wallace, Barbara McClintock, Ernst Caspari, Curt Stern, and occasionally with Max Delbruck, Al Hershey, and Renato Dulbecco, as well as many visiting scientists from Europe, such as Jacques Monod, Francois Jacob, Andre Lwoff, Boris Ephrussi, and Adriano Buzzati-Traverso. Furthermore, he attempted to learn genetics from books and journals, which he supplemented by attending seminars and personal contacts.

With genetics, Mayr lacked the close familiarity that he enjoyed with ornithology, biogeography, and systematic zoology, which formed his basic intellectual foundation. However, as an "outsider" to genetics, Mayr was able to bring a fresh perspective and question the validity of long-established foundations of classical genetics, such as the gene–phenotype correlation. That was very much evident in his opening address to the Cold Spring Harbor symposium "Genetics and Twentieth-Century Darwinism" in 1959. Clearly, he enjoyed asking provocative questions.

In developing his views on the biological species concept and the process of geographic speciation, Mayr was deeply influenced by the German biologists Erwin Stresemann and Bernhard Rensch. Mayr pointed out on more than one occasion the decisive influence of Stresemann and virtually everything in Mayr's 1942 book was "somewhat based on Stresemann's earlier publications" (Haffer 2007, p. 205).

While promoting the acceptance of this species concept with its emphasis on populations and on reproductive isolation (which was earlier emphasized by Dobzhansky), Mayr's concise formulation led to their wider acceptance among biologists. Although these concepts were already established by naturalists in the nineteenth century, Mayr brought them into the twentieth-century biology, with additional evidence and discussion that led to their rapid adoption by new generations of biologists.

Adopting the principles of "new systematics" as applied by Stresemann, Rensch, and himself, Mayr emphasized that the replacement of typological thinking by population thinking is perhaps the greatest conceptual revolution that has taken place in biology. He saw himself as the messenger of this new philosophy of biology, which is holistic in its outlook. He clearly saw the early methods in theoretical population genetics as too reductionist and unrealistic. These, in Mayr's opinion, led to simplistic and nonholistic concepts and conclusions that are not acceptable.

Behind his critique of population genetics lay his long experience in biogeography and his conception of biological speciation, which was totally ignored by the founders of classical population genetics, Fisher, Haldane, and Wright. He criticized the early formulation of each gene acting in isolation of all other genes, which was the foundation for Haldane's and Fisher's early mathematical studies of evolution.

The Causes of Evolution

Mayr's (1992) review of the reprint of Haldane's *The Causes of Evolution* is of interest in this context. Mayr acknowledged that Haldane (1932a) rightly considered genetics as the basis of understanding of evolution. Mayr took into account the papers of Haldane in population genetics that were published for more than thirty years after the publication of *The Causes*. Of special interest is Haldane's (1957b) "cost" estimate of evolution.

Mayr made reference in particular to the arguments presented in support of Haldane's mathematical work by Sewall Wright, James F.

Crow, and Motoo Kimura in the memorial volume *Haldane and Modern Biology* (Dronamraju 1968). And he questioned the validity of the assumptions made by Haldane (1957b) in estimating the "cost of evolution." Mayr's (1992) comments were concerned with two publications of Haldane: his classic *The Causes of Evolution*, which was published in 1932, and a 1959 essay on natural selection.

The primary goal of *The Causes of Evolution* was to synthesize Mendelian genetics and Darwinian evolution. Haldane pointed out that the weakest point in Darwin's argument was the nature of variation and its causes. Haldane therefore took some pains to refute some of Darwin's errors, such as his emphasis on blending inheritance. He pointed out that acceptance of blending inheritance would result in a colossal reduction of variation in every generation and would require an equally colossal production of new genetic variation. When Haldane wrote *The Causes*, it was still necessary to refute all claims of the Lamarckian inheritance of acquired characters. Haldane showed that the results of various experiments that had been interpreted as demonstrating an inheritance of acquired characters had actually been due to unconscious selection. In fact, both Fisher and Haldane did such a great job of refuting soft inheritance that five years later Dobzhansky (1937) devoted very little space to that subject.

Saltationism   Among other topics, the saltationism of the Mendelians was another issue discussed by Haldane. Mayr commented that Haldane displayed an ambivalence, believing in gradual evolution by natural selection on the one hand, and some acceptance of the role of saltational speciation on the other. Generally, however, Haldane (1932a) insisted that selection can almost always overcome any effects of mutation, concluding that "we cannot regard mutation as a cause likely by itself to cause large changes in a species." (p. 60)

Mayr commented that the saltationism of the Mendelians clearly reflected their typological thinking. However, regarding Haldane's thinking, Mayr wrote: "Haldane had gotten rid of this to a large extent. Yet, some of his statements on mutations are still tainted with typological thinking, and even more so are some of his statements on species. For him a species was largely an aggregate of genes. (Mayr 1992)"

And the crux of what troubled Mayr about the early papers of Haldane: "And, like Darwin, he was much more concerned with showing that species are related to each other and that one can derive one species from another, than to show how this occurs. (Mayr 1992)"

With reference to natural selection, Mayr wrote: "I have always admired Haldane for the frankness with which he acknowledged his ignorance about certain problems." Haldane wrote: "We know very little about what is actually selected (Haldane 1932)" and that much of the opposition to natural selection is due to a failure to appreciate the extraordinary subtlety of the principle of natural selection. Mayr pointed out that Haldane's comment is as appropriate today as it was in 1930. Mayr wrote: "Haldane was the most open-minded of the Fisher-Haldane-Wright trio. He perceived aspects of evolution that are often ignored. For instance, he was fully aware of the frequency of neutral or near-neutral genes. (Mayr 1992)"

Genes in Isolation   It is perhaps to Mayr's (1992) surprise that he noted Haldane's caution about thinking of genes in isolation. Mayr (1992) wrote: "Haldane realized that it could lead to deceiving conclusions if one looked at each gene in isolation, because this would fail to reveal synergistic and epistatic interactions." (p. 180) Haldane (1932a) wrote: "It is important to realise that the combination of several genes may give a result quite unlike the mere summation of their effects one at a time... . So selection acting on several characters leads, not merely to novelty, but to novelty of a kind unpredictable with our present scientific knowledge, though probably susceptible of a fairly straightforward biochemical explanation" (p. 96).

Mayr (1992) noted that Haldane was "fully aware of the evolutionary significance of unique events and unique combinations of genes" (p. 180). Haldane (1932a) wrote: "In a great disaster or a great migration the characters of the single survivor are what matters and ... here at least we must admit that mere chance is likely to play a certain part in species formation" (p. 118). Haldane (1932a) discussed other evolutionary topics, such as the role of altruism, but did not develop a full-fledged theory of kin selection.

Evolutionary Problems Neglected by
Haldane and Others

Mayr (1992) noted that Fisher, Haldane, and Wright were mostly preoccupied with changes in gene frequencies in evolution. This is "vertical evolution" in Mayr's sense. Mayr was interested in another aspect of evolution, the "horizontal" evolution, which is concerned with geographical variation, biodiversity, and differences among populations. By extension, it deals with geographic (allopatric) speciation, incipient species, the origin of isolating mechanisms, and the relations between speciation and macroevolution. The emphasis throughout is on populations.

Mayr wrote: "Fisher, Haldane and Wright did not have the background to deal with this aspect of evolution, and almost entirely neglected it in their writings ... It was the major contribution of the naturalists to have brought the evolution of biodiversity into the evolutionary synthesis." Mayr (1992) goes on:

> All three mathematical geneticists treat the species rather typologically. Typological thinking seems also to be the reason why Haldane considers rather tolerantly the saltational speculations of Willis (*Age and Area*) and of De Vries... No mention is made here of populations, of geographic isolation, of the origin of isolating mechanisms, or of any of the other factors that are the normal causative agents of speciation (p. 181).

The Gene: The Target of Selection
(Beanbag Genetics)

Mayr (1992) stated that one consequence of the assumption that the gene is the target of selection (Haldane 1924) is the belief by Haldane (and Fisher) that the efficiency of selection is proportional to the variance of a species, resulting in a rapid evolution in numerous species (Haldane 1932a, p. 132). Yet Haldane realized that the fossil record does not support a belief in such an "accelerated evolution."

Haldane raised the possibility that the breakup of a species into local races could lead to the formation of new species. Mayr noted with some satisfaction that here Haldane touched upon the possibility of geographic speciation.

Speciation   Haldane discussed the importance of small, isolated populations for speciation. All three—Haldane, Fisher, and Wright—believed that in a normal population there is usually an equilibrium, but evolution requires a change or shift in this equilibrium. The question arises as to what can cause such a change. Haldane (1932a) suggested that a change from one stable equilibrium to the other may take place as the result of the isolation of a small unrepresentative group of the population, a temporary change in the environment that alters the relative viability of different types, or in several other ways. He considered such a change to be very important because it is probably the basis of progressive evolution of many organs and functions in higher animals, and of the breakup of one species into several. This is, in fact, similar to Wright's "shifting balance theory." This is one of the few situations where Haldane referred to speciation, but he did not mention any details about the population aspects (see Haldane 1931).

Mayr (1992) noted that although several years later Haldane (1959) came to appreciate the importance of geographic speciation, he still had some reservations. Haldane (1959) wondered: "I think it is very hard to see how the wealth of species swimming and floating in the open sea developed by allopatric speciation" (p. 142). Mayr (1992) responded that Haldane offered no other explanation except quoting William Thorpe (1930), who had promoted sympatric speciation by selection for new host preferences. Mayr (1947) had shown earlier that Thorpe's argument is not at all applicable to pelagic species. Haldane (1959) ends his discussion with an amusing prophecy: "My guess is that, as Mayr's views are now orthodox, they will be violently criticized in twenty years or so, and that most, but not quite all, of these criticisms will prove to be invalid" (p. 143).

Sexual Selection   Mayr (1992) pointed out further that Haldane (1932a) neglected sexual selection, or selection for mating success, a special case of the broader concept of selection for success in social competition: Nowhere did Haldane ask the important question as to the payoff of sexual selection; he focused instead on the sensory issues and aesthetics. However, no satisfactory explanation was offered to account for the neglect of sexual selection between 1900 and the 1970s. Mayr suggested that the assumption that the gene rather than the individual is the target of selection virtually wiped out the distinction between the two kinds of selection.

During the emergence of the modern synthesis the central issue was speciation. The focus of study was on those behavioral aspects which served as isolating mechanisms. Haldane assumed at times that selection is for the benefit of the species and that competition among the members of a species may be injurious for the species. Haldane (1932a) further assumed that it is a fallacy to think that natural selection will always make an organism fitter in its struggle with the environment.

In some species with dense populations, Haldane wrote: "Its members inevitably begin to compete with one another ... [e.g.,] the struggle for mere space which goes on between neighboring plants of closely packed associations." (Haldane 1932a, p. 53) Haldane thought that the results may be biologically advantageous for the individual, but ultimately disastrous for the species. And the "biological effects of competition ... (most likely) render the species as a whole less successful in coping with its environment" and "the special adaptations favored by intraspecific competition divert a certain amount of energy from other functions" (Haldane 1932a, pp. 125–126).

Haldane may have been referring to sexual selection as he knew about the potential incompatibility between sexual selection and natural selection, because he referred later to the "bright colors and songs of many bird species ... probably preserved and enhanced by competition between the members of the species, (although) their value to the species as a whole is dubious" (p. 128). Haldane was skeptical of the reproductive advantages of more extreme sexual dimorphism, such as peacocks with more ornate and longer tails because he thought they might be handicapped by a shorter life span or other disadvantages. They certainly would be slower in getting away from a predator because of the cumbersome tail. Here is an instance of a conflict between natural selection and sexual selection that Haldane had earlier alluded to. He suggested that the possible advantages of more extreme sexual dimorphism can be verified by collecting statistics of the number of offspring left on the one hand, and longevity on the other, of long- and short-tailed peacocks under natural conditions (Haldane 1932a, p. 120).

Mayr (1992) cited the example of African widow birds where the data have clearly demonstrated the reproductive advantage of ornamentation, however, it was not known if such selection for reproductive success had any impact on the life expectation. In his 1959 essay "Natural Selection," Haldane still displayed some skepticism as to the importance of sexual selection. He made the distinction between the two unintended consequences of ornamental characters, to attract the females, but may also intimidate or conquer other males. The distinction between the two is not always clear; for instance, the songs of male birds, and possibly their bright plumage, drive other males away, as well as sometimes attracting females. After citing some examples of sexual selection from *Drosophila,* Haldane (1959) concluded that, as Darwin had predicted, the choice depends on hereditary characters of *both* sexes. But, he added that "it is too early to say whether sexual selection is as important as Darwin believed, or perhaps even more important. There is at least no doubt that it is a reality" (p. 132).

Reductionism   On the subject of reductionism in genetics and the possibility of reducing biology to physics, Haldane (1932a) wrote:

There is a tendency in some quarters to describe the phenomenon with which I have just dealt as "the mechanism of heredity," and to suppose that the introduction of atomism by Mendel has reduced genetics to biophysics. I do not think that this is so. We can, in principle at least, speak of the mechanism of segregation. But the things segregated, the genes, reproduce themselves or are copied at each cell division. And this process of reproduction cannot at present be explained in physico-chemical terms, whatever may be possible in the future…. It is at present irrelevant to genetics whether life is or is not ultimately explicable in terms of physics and chemistry. Hence it is irrelevant to the general argument of this book, which is based on the facts of genetics (p. 33).

Haldane (1932a) foresaw the future possibility of a successful physicochemical analysis of the genetic material, as was ultimately achieved by Watson and Crick. It is interesting that, in his commentary, Mayr (1992) not only did not object to Haldane's discussion of reductionism but appreciated Haldane's honesty about the lack of a molecular explanation at that time.

Mayr's (1992) evaluation of Haldane's (1932a, 1959) views on evolution as enunciated in two publications separated by twenty-seven years is of great interest because his discussion covered a whole range of issues related to beanbag genetics, such as variation, saltationism, genes in isolation, target of selection, speciation, natural selection, sexual selection, and reductionism. To his surprise, Mayr found that Haldane was clearly aware of the pitfalls of looking at genes in isolation, which was one of the key issues in his critique of "beanbag" genetics (Mayr 1959). It appears that the views of Mayr (1992) are not the same as those of Mayr (1959).

Perhaps, Mayr (1992) had the opportunity and time to view the entire views of Haldane in a broader context. Mayr presented contradictory views. He readily admits that Haldane was the most open minded scientist that he had ever known. Or perhaps he began to appreciate that Haldane and Fisher in their early analyses of the mathematical theory of natural selection adopted a method of treating genes in isolation (although not in their later papers) for the sake of mathematical analyses. Although Mayr (1959) included Wright as well in his criticism, he was clearly mistaken. Wright (1931 and later) was the only member of the trio who consistently took into account the impact of epistasis and other genic interactions on evolution.

Whatever the reasons were, in his later years, Mayr (1992) appears to have gained a better understanding of Haldane's contributions to population genetics. He still continued to maintain his disagreements but he seemed to have realized that Haldane's views on studying genes in isolation and other topics were not too different from his own. However, certain important differences remain even today.

The trio of Haldane, Fisher, and Wright paid almost no attention to the process of speciation, a key topic of Mayr's life-long research. Their study of evolution was carried out through mathematical analyses which focused almost exclusively on changes in gene frequencies. Mayr (1959) appears to have not given adequate credit to Haldane, Fisher, and Wright for laying the initial foundation of the "modern synthesis." Their combined contribution to the evolutionary theory was called the first phase of the modern synthesis by Gould (2002). With the exception of Mayr, all other founders of the modern synthesis, Huxley, Dobzhansky, Simpson, and Stebbins, appear to have fully grasped the importance of this first phase that was discussed by Gould (2002).

Beanbag Genetics Today

Crow (2001, 2008, 2009) has written eloquently on the recent and future achievements of beanbag genetics. The application of beanbag genetics did not stop with Haldane, Fisher and Wright. Crow (2001) wrote:

> The gene-pool model is a wonderful, simplifying convention…. The simplicity and power of treating random sampling from a pool of genes as equivalent to random mating of diploid individuals is a bonus…. Ernst Mayr compared this type of gene-pool modelling to sampling from a bag of coloured beans and dismissed it as "beanbag genetics."… As a means of bypassing distracting details to provide useful approximations and novel insights, gene-pool models are here to stay. (p. 771)

Beanbag genetics has kept up with later developments in at least one respect. Drawing colored beans from a bag forces one to attend to random processes, which are so important in the study of molecular evolution. The essential randomness continues even though the models may become more complex. The bean pool may change, but the basic process remains (Crow 2008).

Today, gene action is much better understood. Attention has shifted to the important role of regulation from the earlier emphasis on transcription and translation. Paradoxically, as our understanding of physiological mechanisms in gene action has increased, so has the importance of beanbag genetics. Contrary to Haldane's (1964) complaint that adequate data were not forthcoming to test theoretical models, there is an abundance of data today from molecular genetics.

The unit of observation is the nucleotide that can be observed and studied with precision, replacing the older concept of the gene which was vaguely inferred from the phenotype. Contemporary beanbag genetics includes molecular clocks, nucleotide diversity, coalescence, and DNA-based phylogenetic trees, along with the four major holdovers from classical genetics, mutation, selection, migration, and random drift. Molecular genetics has made it possible to study evolution rates at the nucleotide level. It is also possible today to compare DNA similarities and divergence in diverse species of animals and plants, which were not crossable previously.

Crow (2008) mentioned another instance of a recent application of beanbag genetics: "The 'out of Africa' concept in human ancestry depended on beanbag genetics, since the most important evidence came from nucleotide diversity"(p. 445). The new beanbag genetics includes Kimura's (1983) neutral theory, and, according to Crow (2008), Kimura should be crowned as the "philosopher-King" of the neo-beanbaggists. Molecular studies of vertebrates have revealed an overwhelming fraction of noncoding DNA. Kimura has argued that most evolution in these regions is driven by neutral mutation, with random drift defining the fate of individual mutations.

The neutral theory has supplied the foundation for a molecular clock as well as becoming the accepted null hypothesis for measuring selection. The molecular clock technique is an important tool in molecular systematics, the use of molecular genetics information to determine the correct scientific classification of organisms. Knowledge of approximately constant rate of molecular evolution in particular sets of lineages also facilitates establishing the dates of phylogenetic events.

Last Exchange

The last exchange of letters between Haldane and Mayr were poignant as Haldane knew he was dying of cancer. On September 29, 1964, Haldane wrote to Mayr that his cancer was spreading and he was too weak to continue the correspondence. In his reply of October 21, 1964, which was addressed to Mrs. Haldane (Dr. Helen Spurway), Mayr expressed sympathy and support, mentioning some recollections of Haldane. Mayr remembered that Haldane was the only one of the older geneticists present at an international conference in Pavia who fully appreciated what the younger ones were talking about. Mayr wrote:

"R.A. Fisher disappointed me at the time by simply resisting the new ideas. Haldane accepted them at once (so far as they were sound) and asked meaningful questions as to where to go from here. This flexibility of his mind no doubt is the reason for his many original ideas. His 1957 paper on the cost of natural selection is another documentation of his continuing originality… . Please give Haldane my best wishes. He asked me not to write to him directly and so I am writing to you."

Haldane died of Cancer on December 1, 1964 in India. Many years later, on July 31, 2002, Mayr responded to my request, giving me permission to reproduce his correspondence with Haldane. He wrote further: "I always tell everybody that Haldane was the most brilliant person I ever met in my life; considering his enormous gifts it has always been a puzzle to me that he has not made more great discoveries that would be recorded in a history of biology. I think he had too many interests and never concentrated on any specific problem in evol. biology… . Of course, I feel honored that you plan to publish some of my letters and I herewith authorize you to reprint them or parts of them. Since Haldane's thinking was in some respect so similar to mine, it was a great loss for me that he died so prematurely. I also authorize you to select the portions of my letters which you want to publish.

Don't forget to mention what an extraordinary interest in natural history Haldane had, as I discovered on my excursion with him in Bhubaneshwar."

References

Ali, S. (1985) *The Fall of a Sparrow*. New Delhi: Oxford University Press.

Ayala, F.J. (1977) Nothing in biology makes sense except in the light of evolution. *J. Hered., 68*: 3–10.

Bateson, W. (1902) *Mendel's Principles of Heredity: A Defense*. Cambridge: University Press.

Bateson, W. (1906) Presidential address. Third International Conference on Genetics, Royal Horticultural Society, London.

Bateson, W. (1909) *Mendel's Principles of Heredity*. Cambridge: University Press.

Bateson, W. (1914) Heredity. *Nature, 93*: 635–642.

Behnke, A.R. and R.W. Brauer (1968) Physiologic investigations in diving and inhalation of gases. In: Dronamraju, K.R. (ed) *Haldane and Modern Biology*. Baltimore: Johns Hopkins University Press., pp. 267–276.

Bernstein (1924) Ergebnisse einer biostatistichen zusammenfassenden Betrachtung uber die erblichen Blutstrukturen des Menschen. *Klin. Wschr. 3:* 1495–1497.

Bernstein, F. (1931) Zur grundlegung der vererbung beim menschen. *Z. Indukt. Abstamm. Vererbungsl., 57*: 113–118.

Birch, L.C. (1954) Experiments on the relative abundance of two sibling species. *Aust. J. Zool., 2*: 66–74.

Bock, W.J. (2004) Ernst Mayr at 100—a life inside and outside of ornithology. *Auk, 121*: 637–651.

Bock, W.J. (2005) Ernst Mayr—teacher, mentor, friend. *J. Biosci., 30*: 422–426.

Bock, W.J. (2006) Ernst Walter Mayr. 5 July, 1904–3 February 2005. *Biogr. Mem. Fellows R. Soc. Lond., 52*: 167–187.

Bock, W.J., and Lein, M.R., eds. (2005) Ernst Mayr at 100. Ornithologist and naturalist (with interview and remarks by Ernst Mayr on Video CD-ROM). *Ornithol. Monogr., 58*: 1–109.

Box, J. F., (1978) *R.A. Fisher: The Life of a Scientist*. New York: John Wiley & Sons.

Briggs, G.E. and J.B.S. Haldane (1925) Note on the kinetics of enzyme action. *Biochem. J., 29*: 338–339.

Calmette, A., and Guérin, C. (1924) Vaccination des bovides contre la tuberculose et methode nouvelle de prophylaxie de la tuberculose bovine. *Ann. Inst. Pasteur, 38*: 171.

Carlson, E. (1981) *Genes, Radiation, and Society: The Life and Work of H.J. Muller*. Ithaca, N.Y.: Cornell University Press.

Carlson, E. A. (2004) *Mendel's Legacy: The Origin of Classical Genetics*. Cold Spring Harbor, N.Y: Cold Spring Harbor Laboratory Press.

Carson, H.L. (1980) Cytogenetics and the neo-Darwinian synthesis. In: Mayr, E., and Provine, W. (eds.), *The Evolutionary Synthesis*. Cambridge, MA: Harvard University Press, pp. 86–95.

Chetverikov, S.S. (1926) On certain aspects of the evolutionary process. Engl. trans., *Proc. Amer. Phil. Soc., 105*: 167–195, 1961.

Coyne, J.A., and Orr, H.A. (2004) *Speciation*. Sunderland, MA: Sinauer Associates.

Crow, J.F. (1958) Some possibilities for measuring selection intensities in man. *Human Biol., 30*: 1–13.

Crow, J. F. (1984) The Founders of Population Genetics. In Chakravarthi, A. (Ed) *Human Population Genetics*. New York: Van Nostrand Reinhold.

Crow, J.F. (1990) R.A. Fisher, a centennial view. *Genetics, 124*: 207–211.

Crow, J.F. (1992) Centennial: J. B. S. Haldane, 1892–1964. *Genetics, 130*: 1–6.

Crow, J.F. (1993) Felix Bernstein and the first human marker locus. *Genetics, 133*: 4–7.

Crow, J.F. (1994) Advantages of sexual reproduction. *Dev. Genet.*, 15: 205–213.

Crow, J.F. (2001) The beanbag lives on. *Nature, 409*: 771–773.

Crow, J.F. (2004) Felix Bernstein and the first human marker locus. *Genetics, 133*: 4–7.

Crow, J.F. (2008) Commentary: Haldane and beanbag genetics. *Int. J. Epidemiol., 37*: 442–445.

Crow, J.F. (2009) Mayr, mathematics and the study of evolution. *J. Biol., 8*: 10–13.

Crow, J.F. and M. Kimura (1965) The theory of genetic loads. *Proc. XI Intl. Cong. Genet., 3*: 495–505.

Cuenot, L. (1903) Hypothese sur l'Heredite des couleurs dans les croisements des souris noires, grises et blanches. *C. R. Soc. Biol., 55*: 301–302.

Darwin, C. (1859) *On the Origin of Species by Means of Natural Selection or the Preservation of Favored Races in the Struggle for Life.* London: Murray.

Dawkins, R. (1976) *The Selfish Gene.* New York: Oxford University Press.

Dobzhansky, T. (1933) Geographical variation in lady-beetles. *Amer. Nat, 67*: 97–126.

Dobzhansky, T. (1935) A critique of the species concept in biology. *Philos. Sci. 2*: 344–355.

Dobzhansky, T. (1937) *Genetics and the Origin of Species.* New York: Columbia University Press.

Dobzhansky, T. (1950) Mendelian populations and their evolution. *Amer. Nat., 84*: 401–418.

Dobzhansky, T. (1951) *Genetics and the Origin of Species.* 3rd ed. New York: Columbia University Press.

Dobzhansky, T. (1960) Book review: *Darwin's Biological Work Some Aspects Reconsidered* (ed.) P.R. Bell. *J. Genet., 57*: 167–168.

Dobzhansky, T. (1973) Nothing in biology makes sense except in the light of evolution. *Am. Biol. Teacher, 35*(3): 125–129.

Dronamraju, K.R., ed. (1968) *Haldane and Modern Biology.* Baltimore: Johns Hopkins University Press.

Dronamraju, K.R. (1985) *Haldane: The Life and Work of JBS Haldane with Special Reference to India.* Aberdeen, U.K.: Aberdeen University Press.

Dronamraju, K.R. (1989) *Foundations of Human Genetics.* Springfield, IL: Charles C. Thomas.

Dronamraju, K.R., ed. (1990) *Selected Genetic Papers of JBS Haldane.* New York: Garland.

Dronamraju, K.R. (1992) *The History and Development of Human Genetics. Progress in Different Countries.* Singapore: World Scientific.

Dronamraju, K.R. (1993) *If I Am to Be Remembered: The Life and Work of Julian Huxley with Selected Correspondence.* Singapore: World Scientific. Ltd.

Dronamraju, K.R., ed. (1995) *Haldane's Daedalus Revisited.* Oxford: Oxford University Press.

Dronamraju, K.R., ed. (2004) *Infectious Disease and Host-Pathogen Evolution.* Cambridge: Cambridge University Press.

Dronamraju, K.R., ed. (2006) *Malaria: Genetic and Evolutionary Aspects.* New York: Springer.

Dronamraju, K.R. (2009) *What I Require from Life? Writings on Science and Life from JBS Haldane.* Oxford: Oxford University Press.

Durham, F. (1911) Further experiments on the inheritance of coat color in mice. *J. Genet.,* 1: 159–178.

Emerson, R.A. and E.M. East (1913) The inheritance of quantitative characters in maize. *Bull. Agric. Exper. Sta. Nebr.,* 120 pp.

Ewens, W.J. (2008) Commentary: On Haldane's "defense of beanbag genetics." *Int. J. Epidemiol.,* 37: 447–451.

Felsenstein, J. (2004) *Phylogenies.* Sunderland, MA: Sinauer.

Fisher, R.A. (1911) Mendelism and Biometry. Unpublished.

Fisher, R. A. (1912) On an absolute criterion for fitting frequency curves. *Messenger of Mathematics,* 41: 155–160.

Fisher, R. A. (1915a) Frequency distribution of the values of the correlation coefficient in samples from an indefinitely large population. *Biometrika,* 10: 507–521.

Fisher, R.A. (1915b) The evolution of sexual preference. *Eugenics Rev.,* 7: 184–192.

Fisher, R.A. (1918) The correlation between relatives on the supposition of Mendelian inheritance. *Trans. R. Soc. Edinb.,* 52: 399–433.

Fisher, R.A. (1922a) On the dominance ratio. *Proc. R. Soc. Edinb.,* 42: 321–341.

Fisher, R.A. (1922b) Darwinian evolution by mutations. *Eugenics Rev.,* 14: 31–34.

Fisher (1928) Two further notes on the origin of dominance. *Amer. Nat.,* 62: 571–574.

Fisher, R.A. (1930) *The Genetical Theory of Natural Selection.* Oxford: Clarendon Press.

Fisher, R.A. (1931) The evolution of dominance. *Biol. Rev.,* 6: 345–368.

Fisher, R.A. (1932) The bearing of genetics on theories of evolution. *Sci. Progr.,* 27: 273–287.

Fisher, R.A. (1935a) The detection of linkage with "dominant" abnormalities. *Ann. Eugen.,* 6: 187–201.

Fisher, R.A. (1935b) The detection of linkage with "recessive" abnormalities. *Ann. Eugen.,* 6: 339–351.

Fisher, R.A. (1958) Polymorphism and natural selection. *Bull. Inst. Int. Stat.,* 36: 284–289; reprinted in *J. Ecol.,* 46: 289–293.

Fisher, R.A. (1973) *Statistical Methods and Scientific Inference.* 3rd ed. New York: Hafner Press.

Fisher, R.A., Ford, E.B., and Huxley, J.S. (1939) Taste-testing the anthropoid apes. *Nature, 144*: 750.

Ford E.B. (1940) Polymorphism and taxonomy. In: Huxley J.S. (ed.), *The New Systematics*. Oxford: Oxford University Press.

Ford, E.B. (1975) *Ecological Genetics*, 4th ed. London: Chapman and Hall.

Fox, A.S. (1949) Immunogenetic studies of *Drosophila melanogaster. II.* Interaction between *rb* and *v* loci in production of antigens. *Genetics, 34*: 647–664.

Garrod, A.E. (1902) The incidence of alkaptonuria: A study in chemical individuality. *Lancet, ii: 1616–1620.*

Garrod, A.E. (1909) *Inborn Errors of Metabolism. London: Oxford University Press.*

Gould, S.J. (2002) *The Structure of Evolutionary Theory*. Cambridge, MA: Harvard University Press.

Haffer, J. (2007) *Ornithology, Evolution, and Philosophy: The Life and Science of Ernst Mayr 1904–2005.* Berlin: Springer.

Haldane, J.B.S. (1919) The combination of linkage values, and the calculation of distances between the loci of linked factors. *J. Genet., 8*: 299–309.

Haldane, J.B.S. (1920). Some recent work on heredity. *Trans. Oxford Univ. Jr. Sci. Club*, 1 (3): 3–11.

Haldane, J.B.S. (1922) Sex-ratio and unisexual sterility in hybrid animals. *J. Genet., 12*: 101–109., p

Haldane, J.B.S. (1923) *Daedalus; or, Science and the Future*, London: Chatto and Windus. pp

Haldane, J.B.S. (1924) A mathematical theory of natural and artificial selection. Part I. *Trans. Cambr. Philos. Soc., 23*: 19–41.

Haldane, J.B.S. (1927a) A mathematical theory of natural and artificial selection. Part IV. *Proc. Cambr. Philos. Soc., 23*: 607–615.

Haldane, J.B.S. (1927b) A mathematical theory of natural and artificial selection. Part V. Selection and mutation. *Proc. Cambr. Philos. Soc., 23*: 838–844.

Haldane, J.B.S. (1929) The origin of life. *Rationalist Annu.*, 3. Chap2 and 3.

Haldane, J.B.S. (1930) A note on Fisher's theory of the origin of dominance, and on a correlation between dominance and linkage. *Am. Nat., 64*: 87–90.

Haldane, J.B.S. (1931) A mathematical theory of natural selection. Part VIII. Metastable populations. *Proc. Camb. Phil. Soc., 27*: 137–142.

Haldane, J.B.S. (1932a) *The Causes of Evolution*. London: Longmans Green.[*]

Haldane, J.B.S. (1932b) A method for investigating recessive characters in man. *J. Genet., 25*: 251–255.

Haldane, J.B.S. (1932c) The time of action of genes, and its bearing on some evolutionary problems. *American Naturalist, 66*: 5–24.

Haldane, J.B.S. (1933) The part played by recurrent mutation in evolution. *Am. Nat., 67*: 5–19.

Haldane, J.B.S. (1935) The rate of spontaneous mutation of a human gene. *J. Genet., 31*: 317–326.

Haldane, J.B.S. (1936) The amount of heterozygosis to be expected in an approximately pure line. *J. Genet., 32*: 375–391.

Haldane, J.B.S. (1937a) The effect of variation on fitness. *Am. Nat., 71*: 337–349.

Haldane, J.B.S. (1937b) *My Friend, Mr. Leaky*. London: The Cresset Press.

Haldane, J.B.S (1939) The theory of the evolution of dominance. *J. Genet., 37*: 365–74.

Haldane, J.B.S. (1940) The blood group frequencies of European peoples and racial origins. *Human Biology, 12*: 457–480.

Haldane, J.B.S. (1941) The relative importance of principal and modifying genes in determining some human diseases. *J. Genet., 41*: 149–157.

Haldane, J.B.S. (1942) Selection against heterozygosis in man. *Ann. Eugenics, 11*: 333–340.

Haldane, J.B.S. (1947a) The mutation rate of the gene for haemophilia, and its segregation ratios in males and females. *Ann. Eugenics, 13*: 262–271.

Haldane, JBS (1948) The theory of a cline. *J. Genet; 48*: 277.

Haldane, J.B.S. (1949a) The rate of mutation of human genes. Proceedings of the Eighth International Congress of Genetics. *Hereditas, 35*: 267–273.

Haldane, J.B.S. (1949b) Disease and evolution. *Ricerca Sci., 19*: 2–11.

Haldane, J.B.S. (1949c) Suggestions as to quantitative measurement of rates of evolution. *Evolution, 3*: 51–56.

Haldane, J.B.S. (1954) The statics of evolution. In: Huxley, J.S., Hardy, A.C., and Ford, E.B. (eds.), *Evolution As a Process*. London: Allen and Unwin, p. 109.

[*] All references to Haldane's *The Causes of Evolution* are from the 1990 edition, Princeton University Press, Princeton, N.J.

Haldane, J.B.S. (1955) Population genetics. *New Biology, 18*: 34–51.

Haldane, J.B.S. (1956) The detection of autosomal lethals in mice induced by mutagenic agents. *J. Genet., 54*: 327–342.

Haldane, J.B.S. (1957a) The detection of sublethal recessives by the use of linked marker genes in the mouse (appendix to T.C. Carter's paper). *J. Genet., 55*: 596–597.

Haldane, J.B.S. (1957b) The cost of natural selection. *J. Genet., 55*: 511–524.

Haldane, J.B.S. (1959) Natural selection. In: P.R. Bell (ed.), *Darwin's Biological Work: Some Aspects Reconsidered*. Cambridge: Cambridge University Press, pp. 101–149.

Haldane, J.B.S. (1960) The interpretation of Carter's results on induction of recessive lethals in mice. *J. Genet., 57*: 131–136.

Haldane, J.B.S. (1964) A defense of beanbag genetics. *Perspect. Biol. Med., 7*: 343–359.

Haldane, J.B.S., with A.D. Sprunt and N.M. Haldane (1915) Reduplication in mice. *J. Genet., 5*: 133–135.

Haldane, J. B. S. and L. S. Penrose, (1935) Mutation rates in man. *Nature, 135*: 907–908.

Haldane, J.B.S. and J. Bell (1937) The linkage between the genes for Colour-Blindness and Haemophilia in man. *Proc. Roy. Soc. Lond.,B,* 123: 119–150.

Haldane, J.B.S. and C.A.B. Smith (1947) A new estimate of the linkage between the genes for colour-blindness and haemophilia in man. *Ann. Eugen., 14*: 10–31.

Haldane, J.B.S. and S.D. Jayakar (1963) Polymorphism due to selection of varying direction. *J. Genet., 58*: 237–242.

Hamilton, W.D. (1964) The genetic evolution of social behavior. *J. Theor. Biol., 7*: 1–52.

Hammond, A. (2009) J.B.S. Haldane. holism, and synthesis in evolution. In: Cain, J., and Ruse, M. (eds.), *Descended from Darwin: Insights into the History of Evolutionary Studies, 1900–1970*. Philadelphia: American Philosophical Society.

Hardy, G.H. (1908) Mendelian proportions in a mixed population. *Science, 28*: 49–50.

Hill, A.V.S., *et al.* (1991) Common West African HLA antigens are associated with protection from severe malaria. *Nature, 352*: 595–600.

Hogben, L.T. (1931) The genetic analysis of familial traits. 1. Single gene substitutions. *J. Genet.*, *25*: 97–112.

Huxley, J.S. (1942) *Evolution: The Modern Synthesis.* New York: Harper.

Huxley, J.S., Hardy, A.C., and Ford, E.B., eds. (1954) *Evolution as a Process.* London: Allen and Unwin.

Irwin, M.R. (1947) Immunogenetics. *Adv. In Genetics, 1*: 133–159.

Johannsen, W. (1909) *Elemente der exakten Erblichkeitslehre.* Jena: Fischer.

Karn, M.N. and L.S. Penrose (1951) Birth weight and gestation time in relation to maternal age, parity and infant survival. *Ann. Eugen., 16*: 147–164.

Kermack, K.A. (1954) A biometrical study of *Micraster corenguinum* and *M. (Isomicraster) senonensis. Philos. Trans. R. Soc. Lond. B, 237*: 375–428.

Kettlewell, H.B.D. (1956) Further selection experiments on industrial melanism in the Lepidoptera. *Heredity, 10*: 287–303.

Kimura, M. (1960) Optimum mutation rate and degree of dominance as determined by the principle of minimum genetic load. *J. Genet., 57*: 21–34.

Kimura, M. (1962) On the probability of fixation of mutant genes in a population. *Genetics, 47*: 713–719.

Kimura, M. (1964) Diffusion models in population genetics. *Jour. App. Probability, 1*: 177–232.

Kimura, M. (1968) Haldane's contributions to the mathematical theories of evolution and population genetics. In: Dronamraju, K.R. (ed.), *Haldane and Modern Biology.* Baltimore: Johns Hopkins University Press, pp. 133–140.

Kimura, M. (1983) *The Neutral Theory of Molecular Evolution.* Cambridge: Cambridge University Press.

Kimura, M. and T. Maruyama (1966) The mutational load with epistatic gene interactions in fitness. *Genetics, 54*: 1337–51.

Kuhn, T.S. (1962) *The Structure of Scientific Revolutions.* Chicago: University of Chicago Press.

Lack, D. (1947) *Darwin's Finches.* Cambridge, MA: Cambridge University Press.

Lerner, I.M. (1954) *Genetic Homeostasis.* New York: Wiley.

Lush, J.L. (1951) Genetics and animal breeding. In: Dunn, L.C. (ed.), *Genetics in the Twentieth Century*, New York: Macmillan.

MacArthur, R.H., and Wilson, E.O. (1967) *The Theory of Island Biogeography.* Princeton, NJ: Princeton University Press.

Malecot, G. (1952) Les Processus stochastiques et la methode des functions generatices on caracteristiques. *Publ. Inst. Stat. Univ. Paris 1,* Fase. 3, 1–16.

Mather, K. (1941) Variation and selection of polygenic characters. *J. Genet.*, *41*: 159–193.

Mather, K. (1953) The genetical structure of populations. *Symp. Soc. Exp. Biol.*, 7: 66–95.

Maynard Smith, J. (1958) *The Theory of Evolution*. London: Penguin Books.

Maynard Smith, J. (1992) Byte-sized evolution. *Nature, 355*: 772–773.

Mayr, E. (1926) Die Ausbreitung des Girlitz *(Serinus canaria serinus L)*. Ein *Beitrag zur Tiergeographie. J. Ornithol*, 74: 571–671. (Ph.D. thesis).

Mayr, E. (1940) Speciation phenomena in birds. *Am. Nat.*, 74: 249–278.

Mayr, E. (1941) *List of New Guinea Birds. A Systematic and Faunal List of the Birds of New Guinea and Adjacent Islands*. New York: American Museum of Natural History.

Mayr, E. (1942) *Systematics and the Origin of Species*. New York: Columbia University Press.

Mayr, E. (1944) The birds of Timor and Sumba. *Bull. Am. Mus. Nat. Hist., 83*: 123–194.

Mayr, E (1947) Ecological factors in speciation. *Evolution, 1*: 263–288.

Mayr, E. (1954) Change of genetic environment and evolution. In: Huxley, J.S., Hardy, A.C., and Ford, E.B. (eds.), *Evolution As a Process*. London: Allen and Unwin, pp. 157–180.

Mayr, E. (1955) Integration of genotypes: Synthesis. *Cold Spring Harbor Symp. Quant. Biol., 20*: 327–333.

Mayr, E. (1959) Where are we? *Cold Spring Harbor Symp. Quant. Biol., 24*: 1–14.

Mayr, E. (1963) *Animal Species and Evolution*. Cambridge, MA: Harvard University Press.

Mayr, E. (1970) *Populations, Species and Evolution*. Cambridge, MA: Harvard University Press.

Mayr, E. (1976) *Evolution and the Diversity of Life*. Cambridge, MA: Harvard University Press.

Mayr, E. (1980) The evolutionary synthesis. In: Mayr, E., and Provine, W. (eds.), *Perspectives on the Unification of Biology*, Cambridge, MA: Harvard University Press.

Mayr, E. (1981) Evolutionary Biology. In: W. Shropshire, Jr., (ed.): *The Joys of Research.*, Washington, D.C.: Smithsonian Institution Press.

Mayr, E. (1982) *The Growth of Biological Thought: Diversity, Evolution, and Inheritance*. Cambridge, MA: Harvard University Press.

Mayr, E. (1985) Darwin's five theories of evolution. In: Kohn, D. (ed.), *The Darwinian Heritage*. Princeton, NJ: Princeton University Press, pp. 755–772.

Mayr, E. (1988) *Toward a New Philosophy of Biology: Observations of an Evolutionist*. Cambridge, MA: Harvard University Press.

Mayr, E. (1991) *One Long Argument. Charles Darwin and the Genesis of Modern Evolutionary Thought*. Cambridge, MA: Harvard University Press.

Mayr, E. (1992) Haldane's *Causes of Evolution* after 60 years. *Quart. Rev. Biol.*, 67: 175–186.

Mayr, E. (2001) *What Evolution is*. New York: Basic Books.

Mayr, E. (2004) *What Makes Biology Unique?* Cambridge: Cambridge University Press.

Mayr, E., and Diamond, J. (2001) *The Birds of Northern Melanesia: Speciation, Ecology and Biogeography*. Oxford: Oxford University Press.

Mendel, G. (1866) Versuche uber Pflanzenhybriden. *Verh. Naturforsch. Ver. Brunn, 4*: 3–47.

Morgan, T.H., A.H. Sturtevant, H.J. Muller, and C.B. Bridges (1915) *The Mechanism of Mendelian Heredity*. New York: Henry Holt.

Morton, N.E. (2008) Commentary: Growth of beanbag genetics. *Int. J. Epidemiol., 37*: 445–446.

Muller, H.J. (1950) Our load of mutations. *Am. J. Hum. Genet., 2*: 111–176.

Muntzing, A.(1932) Cytogenetic investigations on the synthetic *Galeopsis tetrahit. Hereditas, 16*: 105–154.

Neel, J.V., and Valentine, W.N. (1947) Further studies on the genetics of thalassemia. *Genetics, 32*: 38–63.

Nilsson-Ehle, H. (1909) Kreuzungsuntersuchungen an Hafer und Weizen. *Lunds Universit. Arsskr. N.F., 5*: 2: 1–122.

Orr, H.A. (2005) The genetic basis of reproductive isolation: insights from Drosophila. *Proc. Natl. Acad. Sci. U.S.A., 102*: 6522–6526.

Pennisi, E. (2009) Evolution: Modernizing the modern synthesis. *Science, 321*: 196–197.

Penrose, L.S. (1935) Mutation rates in man, *Nature*, 135: 907–912.

Provine, W.B. (1986) *Sewall Wright and Evolutionary Biology*, Chicago: University of Chicago Press.

Punnett, R.C. (1915) *Mimicry in Butterflies*. Cambridge: Cambridge University Press.

Rendel, J.M. (1945) Genetics and cytology of *Drosophila subobscura*. II. Normal and selective matings in *Drosophila subobscura. J. Genet., 46*: 287–302.

Rensch, B. (1929) *Das Prinzip geographischer Rassenkreise und das Problem der Artbildung* [*The Principle of Polytypic Species and the Problem of Speciation*]. Berlin: Borntraeger.

Rensch, B. (1959) *Evolution above the Species Level.* New York: Columbia University Press.

Salisbury, E. (1942) *The Reproductive Capacity of Plants.* London: Bell.

Sarkar, S. (1992) Haldane and the emergence of theoretical population genetics, 1924–1932. *J. Genet., 71*: 73–79.

Schmalhausen, I.I. (1949) *Factors of Evolution.* Chicago: University of Chicago Press.

Sheppard (1951) Fluctuations in the selective values of certain phenotypes in the polymorphic land snail *Cepaea nemoralis* (L). *Heredity, 5:* 125–134.

Simpson, G.G. (1944) *Tempo and Mode in Evolution.* New York: Columbia University Press.

Simpson, G.G. (1950) *The Meaning of Evolution.* New Haven, CT: Yale University Press.

Singh, R.S. (2003) Darwin to DNA, molecules to morphology: the end of classical population genetics and the road ahead. *Genome, 46*: 938–942.

Sober, E. (1984) *The Nature of Selection: Evolutionary Theory in Philosophical Focus.* Cambridge, MA: MIT Press.

Spurway, H. (1949) Remarks on Vavilov's law of homologous variation. *Ric. Sci., 19*(Suppl.): 3–9.

Stebbins, G.L. (1950) *Variation and Evolution in Plants.* New York: Columbia University Press.

Stresemann, E. (1939) "Zoogeography" *in* "Die Vogel von Celebes." *Journal fur Ornithologie, 87*: 312–425.

Sturtevant, A.H. (1965) *A History of Genetics.* New York: Harper and Row.

Todd, A. (1930) Printed from Haldane, J. B.S (1932) The *Causes of Evolution*, Reprinted 1990, Princeton University Press, Princeton, NJ; P.51.

Van Valen, L. (1963) Haldane's dilemma, evolutionary rates and heterosis. *Amer. Nat., 97*: 185–190.

Vavilov, N.I. (1922) The law of homologous series in variation. *Journal of Genetics. 12:* 47–89.

Waddington, C.H. (1953) Epigenetics and evolution. In: Brown, R., and Danielli, J.F. (eds.), *Evolution* (SEB Symposium VII). Cambridge: Cambridge University Press, pp. 186–199.

Waddington, C.H. (1957) *The Strategy of the Genes*. London: Allen and Unwin.

Weatherall, D.S. (2004) J.B.S. Haldane and the malaria hypothesis. In: Dronamraju, K.R. (ed.), *Infectious Disease and Host-Pathogen Evolution*. New York: Cambridge University Press, pp. 18–36.

Weinberg, W. (1908) Über den Nachweis der Verebung beim Menschen. *Verein Vaterl. Naturk. Württemb., 64*: 368–82.

Weldon, W.F.R. (1901) A first study of natural selection in *Clausilia laminata* (*Montagu*). *Biometrika, 1*: 109–124.

Wilkinson, G.S., Kahler, H., and Baker, R.W. (1998) Evolution of female mate preferences in stalk-eyed flies. *Behav. Ecol. 9*: 525–533.

Willis, J.C. (1922) *Age and Area*. Cambridge: Cambridge University Press.

Wright, S. (1921) Systems of mating. *Genetics, 6*: 111–128.

Wright, S. (1922) Coefficients of inbreeding and relationship. *Am. Nat. 56*: 330–338.

Wright, S. (1929) Evolution in a Mendelian population. *Anat. Rec., 44*: 287.

Wright, S. (1931) Evolution in Mendelian populations. *Genetics, 16*: 97–159.

Wright, S. (1968a) Contributions to genetics. In: Dronamraju, K.R. (ed.), *Haldane and Modern Biology*. Baltimore: Johns Hopkins University Press, pp. 1–12.

Wright, S. (1968b) *Genetic and Biometric Foundations*. Evolution and the Genetics of Populations, vol. 1. Chicago: University of Chicago Press.

Appendix

30   Mayr to Haldane, (reply), one page—March 16,
         1961   243

31   Haldane to Mayr, one page—April 19, 1961
         (Haldane's statement for the Kimber Medal ceremony
         at the National Academy of Sciences in Washington, D.C.
         is enclosed).   244

32   Mayr to Haldane, one page—April 26, 1961   245

33   Haldane to Mayr, two pages—February 19, 1962   248

34   Mayr to Haldane, one page—March 1, 1962   250

35   Mayr to Haltdane and Helen, one page—November 19,
         1962   251

36   Mayr to Haldane, one page—February 25, 1963   252

37   Haldane to Mayr, one page—April 6, 1963 (Mayr's
         "Animal Species and Evolution"   253

38   Haldane to Mayr, two pages—May 8, 1963
         (Beanbag genetics)   254

39   Haldane to Mayr, one page—June 3, 1963
         (defense of beanbag genetics)   256

40   Mayr to Haldane, two pages—June 18, 1963   257

41   Haldane to Mayr, (reply) two pages—June 26,
         1963   259

(39 and 40 discuss Haldane's article on the origin of lactation)

42   Dr. Helen Spurway (Mrs. Haldane) to Mayr,
         one page—November 4, 1963   261

43   Mayr to Haldane, one page—April 20, 1964 (Haldane's
         cancer and poem)   262

44   Haldane to Mayr, two pages—May 2, 1964   263

45   Mayr to Haldane, two pages—June 3, 1964   265

(44 and 45 discuss Haldane's election to the U.S. National
Academy of Sciences)

46   Haldane to Mayr, one page—September 29, 1964   267

47   Mayr to Dr. Helen Spurway (Mrs. Haldane),
         one page—October 21, 1964   268

48   Mayr to Dronamraju (Author), one page—July 31,
         2002   269

Haldane died on December 1, 1964.

Dear Dr. Mayr

Thank you for your footnote.
It always hard to know how much maths
use. I rather took it for granted that
coefficient of variation is nearly the
standard deviation of the natural logarithms of
"scale" was as a result one could take as
... Clearly I or my secretary put
"mens for *solœnsis* at one point. The
key thing is the lowness of *H. solœnsis'*
ed. I don't think "pœlaeo" is any better
"pœleo". Why ... A Darwin is a
te of 10^{-6}, so 4×10^{-8} is 40 millidarwins

undated

ROYAL INSTITUTE OF BRITISH ARCHITECTS

ARCHITECTURAL SCIENCE BOARD LECTURE
at 66 PORTLAND PLACE, W.1, on

Tuesday, February 15th, 1949, at 6 p.m.
FORCE AND FORM
The aesthetics of stress distribution
by Felix J. Samuely, B.Sc., Assoc.M.Inst.C.E., M.I.Struct.E., M.I.W.
To be illustrated with lantern slides.

SYNOPSIS

1. Introduction to the three basic types of structure: beams and columns; arches and frames; vaults and domes.
2. The shape of structural members as dictated by external forces, internal stresses, available materials and technique.
3. The general attitude towards the architectural expression of stresses: showing the structure; emphasizing the structure; hiding the structure.
4. Means of making the stresses visible and of concealing them, exemplified by arches and beams.
5. Special reference to structural shapes in modern materials, such as riveted and welded steel, cast in-situ and precast concrete, aluminium, etc.
6. Latticed and half-latticed construction; their frequent lack of expression, and the means of overcoming this.

C. D. SPRAGG,
Secretary, R.I.B.A.

Light refreshments will be provided from **5.5** p.m. to **5.55** p.m.

undated

UNIVERSITY COLLEGE LONDON
DEPARTMENT OF BIOMETRY   GOWER STREET, W.C.1

EUSton 4400

Professor J. B. S. HALDANE

Dear Mayr

I hope you will spend at least a day here. The following should interest you. My wife has crossed the following subspecies of Triturus cristatus : cristatus, London, carnifex, Naples, danubialis Budapest, karelinii, Bakis. All F₁ s have a meiosis sufficiently normal to show inversions and translocations (Callan). Only about 20% of back-crosses get them to metamorphosis. But they show segregation in a big way. cristatus and carnifex differ by at least 3, perhaps 8 genes. She would like you to see them. Gruneberg is doing skeletal variation in detail in mice (pure lines and wild populations) and getting very striking results.

Our other stuff is of less general interest to you, I think

We shall of course be delighted if you can stay with us, as Prof Norbert

Wiener has recently done. But it is only fair to warn you that

1. Our flat is full of cuts, and stinks vilely

2. We are apt to work till midnight, so you would have to look after yourself to some extent.

But bed and breakfast are yours for the asking, and we would like to have you very much if only for a few days. We have what may be a new idea about speciation, which I will discuss. Here is (probably) an old one in tabloid form. "The genetically determined differences between members of a population tend to be due to genes or genoids (i.e. inversions etc behaving as units in meiosis) whose heterozygote is fitter than either homozygote; differences between members of subspecies species etc tend to be due to genes or genoids whose heterozygote is less fit". This would include all translocations, and some antigenic differences

My wife joins me in hoping you will spend at least a few nights with us (unless your F.B.I. would object).

                                        Yours sincerely,
                                        J.B.S. Haldane

Dear Mayr

Thank you for your letter. Fresnaye way _C. oryzae_ in England is pretty unknown, and he is not prepared to do extensive trapping of small rodents in foreign countries, especially in Switzerland and Sweden when a long he were arrested for killing game without a licence!

Meanwhile could you be so good as to let me have a reprint of Mayr, and to discuss on the ducks & geese (Wilson Bulletin abt. 1945)? Scott has got a number of interspecific hybrids and my wife helps to get some of them suitably mated. Several intergeneric hybrids made there, e.g. _Meas_ – _Anas_, but they are more likely to be sterile.

Yours sincerely

undated

EVOLUTION

International Journal of Organic Evolution

PUBLISHED BY

The Society for the Study of Evolution

ERNST MAYR, *Editor*
THE AMERICAN MUSEUM OF NATURAL HISTORY
CENTRAL PARK WEST AT 79TH STREET
NEW YORK 24, N. Y.

August 6, 1947

Dr. J. B. S. Haldane
University College
Gower Street
London, W. C. 1, England

Dear Dr. Haldane,

The first double issue of EVOLUTION has no doubt reached you by now. How did you like it? I would greatly appreciate getting comments from the Associate Editors on style and contents, as well as suggestions concerning possible changes and innovations. Particularly welcome are suggestions of suitable manuscripts. It is planned to keep the scope of the journal as broad as possible. The disproportionate representation of <u>Drosophila</u> in the first two issues is due to a shortage of manuscripts in other fields. This will change since a considerable number of evolutionary papers from the fields of anthropology, paleontology, and taxonomy have been promised to the Editor.

The subtitle of the journal "<u>International</u> Journal of Organic Evolution" is meant seriously. Every effort is being made to make the journal truly international and your cooperation in this matter is particularly requested.

The next issue (approximately 105 pages) will be published about October 1, the last issue (50-60 pages) on about December 10. Next year there will be four issues.

The suggestion has been made to include in the journal a subject-classified abstract section (as, for example, in the JOURNAL OF ANIMAL ECOLOGY). Even though it is doubtful whether this can be undertaken, it would be valuable to me to have your reaction to this plan.

Any other suggestion that you may have will be greatly appreciated.

Yours sincerely

E. Mayr

Ernst Mayr

EM:sp

23rd August 1947.

Dear Mayr,

Thank you very much for the first number of
EVOLUTION. I should very much like to join the Society,
but at the present time I do not know whether this is
going to be possible owing to financial restrictions.
The membership is so cheap that I certainly want to,
but it is not clear how I shall manage it. With regard
to the first number, I quite agree that <u>Drosophila</u> is
over-represented. I am glad that this is going to change.
I hope that to give the thing balance you will occasionally
produce a mathematical paper, although I know they are apt
to be rather abstract.

We are doing a certain number of things here
which might interest you. We have a number of amphibean
hybrids between rather doubtful species, and if we can
get another generation, they will tell us something worth
knowing.

I am busy measuring sea-urchins, partly because
nobody has done much biometry on echinoderms, but mainly
to give a background to my colleague Kermak, who is working
on <u>Micraster</u> from the Cretaceous.

I will let you know further about subscriptions
as soon as we know where we are. However, at present the
Banks are hardly able to cope with the situation.

By the way, I should imagine that your papers
are full of very misleading reports about the situation
here. The average person here is not hit by this financial
crisis. We have to eat a bit more simply, but we are a long
way from hunger. Our main trouble is really shortage of
buildings, and you need not suppose that there is likely
to be a political crisis, or any of the various things
you are probably told. On the contrary, I should think
the effects of our ceasing to borrow will be worse for
the U.S. than for ourselves.

Yours sincerely,

Ernst Mayr,
Editor, EVOLUTION,
American Museum of Nat.History.

ᴇⱽᴼᴸᵁᵀᴵᴼᴺ

International Journal of Organic Evolution

PUBLISHED BY

The Society for the Study of Evolution

ERNST MAYR, *Editor*
THE AMERICAN MUSEUM OF NATURAL HISTORY
CENTRAL PARK WEST AT 79TH STREET
NEW YORK 24, N. Y.

September 12, 1947

Professor J. B. S. Haldane
University College
Gower Street, W. C. 1
London, England

Dear Professor Haldane,

Thank you for your kind letter and the suggestions for EVOLUTION.
I am most anxious to include some mathematical papers and would be
only too happy to print a contribution of yours. Perhaps your
work on Echinoderms may result in a suitable paper. Indidentally,
I was much pleased to learn that you are working on this group of
animals since I myself have done some exploratory work on sea-
urchins after coming back from the Princeton meeting. I felt like
you that there was no other group of marine animals that offers a
similar wealth of quantitative characters. Also, we know nothing
so far on speciation in marine animals and it seems like an excellent group to study. It will take another year or two before
I will be able to get into this work seriously, but I have started
to analyze Mortensen's monograph to see which genera would be most
suitable for my purpose.

Don't worry too much about subscription to EVOLUTION. We shall send
you the first volume anyhow because you are a member of the Editorial
Board. If you could prepare a contribution we would feel more than
amply repaid. I have written to a number of Russian biologists
and have had several promises of manuscripts. Among mathematical
problems the one that interests me most at the present time is the
balance between selection pressure and gene flow between neighboring
populations. Is there any way of determining mathematically how low
gene flow has to drop to permit a break to develop between two
neighboring populations that are subjected to strong and very different selection pressures? I realize, of course, that only a
general answer can be given to this problem, but it appears to me
that there are some aspects that have not yet been fully treated in
the papers of yourself, Fisher, and Sewall Wright. I also have the
feeling that certain aspects of the conflict between intra-population
selection and inter-population selection have not yet received
adequate mathematical attention. What I have in mind can, for example,

be illustrated by the birds of paradise. The most conspicuous males
are apparently most successful in arousing a state of physiological
readiness in the females. In this way they leave more offspring than
other males, but at the same time the species becomes more and more
conspicuous in the male sex and in consequence also more and more
vulnerable to predation. If it weren't for the fact that predators
are virtually absent in New Guinea, I am sure the bizarre birds of
paradise would have never evolved. There is, thus, a balance between
two antagonistic selective trends. What helps the individual within
the species is harmful to the species as a whole. Such developments
as the Irish Elk, etc. might well fit into the same scheme.

As far as fossil material is concerned, I have always been interested
in the question whether the intra-population variability is greater
in rapidly-evolving lines than in slow ones, and vice versa. Perhaps
Micraster material will shed some light on this question. Unfortunately, most paleontological material is not collected sufficiently
carefully to guarantee homogeneity of samples.

Whatever your contribution to EVOLUTION will be I am looking forward
to it with great anticipation.

With the very best regards, also to Mrs. Haldane

                                        Yours sincerely

                                        Ernst Mayr

EM:sp   Ernst Mayr

International Journal of Organic Evolution

PUBLISHED BY

The Society for the Study of Evolution

ERNST MAYR, *Editor*
THE AMERICAN MUSEUM OF NATURAL HISTORY
CENTRAL PARK WEST AT 79TH STREET
NEW YORK 24, N. Y.

November 16, 1948

Dr. J. B. S. Haldane
Department of Biometry
University College
London, England

Dear Dr. Haldane,

I shall be very glad to publish you paper entitled, "Suggestions
as to the Quantitative Measurement of Rates." If it is all right
with you I shall add a few subtitles as, for example, on page 5
Evolutionary Rates in Horses, on page 8 Evolutionary Rates in
Hominids, and on page 10 Standards of Evolutionary Change.

You may be interested in the comments made by one of the readers
of your paper:

"I have only four comments to make, and only the first of these
is possibly important:

"1. I am not quite satisfied that measurement of rates in standard deviations is particularly enlightening or, at least, is the
best standard. It certainly complicates interpretation by introducing hidden factors not always pertinent to the point of study.
It measures population change relative to population variation, and
the relationship is often unduly complex. For instance, if an
evolving population becomes half as variable (and this occurs in
some samples, at least) this rate automatically becomes twice as
fast even if actually the rate of observed change in the measured
variate per year remains absolutely constant! Change in terms of
standard deviations have a special interest, surely, but they are
hard to interpret and, I think, unsuitable as a basic standard for
evolutionary rates.

"2. The year estimates on p. 6 are, as Haldane says, perhaps not
more than 25 per cent out, but he has probably relied on Simpson's
1944 (really 1942) figures which are, I think, just about 25 per
cent too small for Tertiary time.

"3. I wish on his even more dubious generation rates (as on p. 7)
he had made some comment on Simpson's suggestion (independently
made and more strongly emphasized by Zeuner) that length of generation
seems in fact, to have surprisingly little influence on year rates

of evolution.

"4. He gives the impression on p. 1 that Simpson tended, at least, to exclude any but generic estimates and on p. 2 that use of measured changes is a new idea here. Actually, of course, Simpson (like many before him) used these too, and stressed them adequately, I think. Later he, himself, uses some of Simpson's data of this sort. Incidentally, I think he underestimates the validity of generic comparisons in well-worked groups and their value as giving an over-all change rate never obtainable from comparison of single, measurable variates. The inaccuracy is, surely, no greater than in some of the numerical estimates in this paper.

"-- Not worth modifying the paper to meet these comments, however."

Yours sincerely

E. Mayr

Ernst Mayr

EM:sp

EVOLUTION

International Journal of Organic Evolution

PUBLISHED BY

The Society for the Study of Evolution

ERNST MAYR, *Editor*
THE AMERICAN MUSEUM OF NATURAL HISTORY
CENTRAL PARK WEST AT 79TH STREET
NEW YORK 24, N. Y.

February 1, 1949

Dr. J. B. S. Haldane
Department of Biometry
University College
London, England

Dear Dr. Haldane,

I am enclosing the proof of your paper which I hope you can return to me within three days after receipt. Your paper will be included in the March issue of EVOLUTION. You will note that I added a footnote giving a little more detail on one of your calculations. If you approve of it you might want to cancel the "Ed." I would suggest that you adopt throughout the spelling "paleontologist" which, although less correct, is the customary spelling in this country. On galley 56 you seem to be using <u>soloensis</u> and <u>neanderthalensis</u> for the same skulls. Am I interpreting this correctly? I am retaining your manuscript here since I presume that you have a copy in London. When you return the proof to me you may want to clip off the margin except where you have made corrections. (*To save postage*)

I am enclosing a blank on which to order reprints. If you should find it difficult to get the necessary dollars for this purchase, I am sure we could arrange it by my ordering some British books through you. I shall be glad to do whatever is most convenient to you.

With best regards, also to Mrs. Haldane

Yours sincerely

Ernst Mayr

Ernst Mayr

EM:sp
Encl.

7th February 1949.

Dear Dr. Mayr,

　　　　　Thank you for your footnote. It is always hard
to know how much maths to give. I rather took it for
granted that "A coefficient of variation is nearly the
standard deviation of the natural logarithms of a variate"
was a result one could take as known. Clearly I or my
secretary put <u>sapiens</u> for <u>soloensis</u> at one point. The
striking thing is the lowness of <u>H. soloensis</u>' head.
I don't think "palaeo-" is any better than "paleo"-
A Darwin is a rate of 10^{-6} , so 4×10^{-8} is 40 millidarwins.

Yours sincerely,

Dr. Ernst Mayr,
American Museum of Natural History,
Central Park West at 79th St.,
New Yrok 24, N.Y.

28th February 1950.

Dear Dr. Lorenz,

      I am somewhat worried by the question of sympatric speciation in higher animals. This demands that a change due to a single gene substitution should produce animals which will mate with one another but not with the original form. Clearly a change in epigamic characters will not in general do this, even if it affects both sexes, for it will not be accompanied by a change in innate releasing mechanisms. We know from Rendel's work that the mutant yellow in <u>Drosophila subobscura</u> makes normal o s reject yellow o s but yellow o s do not reject normal o s. The sensory basis is unknown. But it only constitutes a one way barrier, since yellow o s do not reject normal o s.

      Possible changes which would have the suggested effect (apart from such changes as change in time of emergence in insects) are

1. Changes in odour. The mutant and the normal might each be accustomed to its own odour, and find that of the other type repulsive. This implies that the odour is not a specific mating stimulus produced by one sex.

2. (Suggestion of Spurway). Changes in innate courtship behaviour where courtship is symmetrical, as in Cochlid fishes. Especially a change in timing might be effective. There is strong evidence suggesting sympatric speciation in African lake Cochlids.

3. Changes in visual characters in birds such as <u>Anser</u> where families do not break up on migration, and imprinting may be important. E.g. if two parents were heterozygous for the same recessive colour gene, the only surviving o and o among their offspring might both be recessives. If so could each become so "imprinted" with the other that they would probably mate, and their offspring which would all resemble them, would tend to mate together, thus starting a new subspecies which might later become a species?

      I should be very glad to have your opinion on the possibility of these and other changes. In plants, the possibilities are much greater, owing to the occurrence of

P.T.O.

- 2 -

self-fertilization, apomixis and polyploidy. Of course
changes can occur within an animal species which lead to
partial sterility when the new type mates with the original
one. But unless there is some psychological cause of
intense assortative mating those will only lead to the
rapid extinction of the new type. It is also not sufficient
that the new type should not evoke mating responses in the
old type. It must also evoke them preferentially in the
other sex of the new type.

                    Yours sincerely,

THE AMERICAN MUSEUM OF NATURAL HISTORY

CENTRAL PARK WEST AT 79TH STREET

NEW YORK 24, N. Y.

March 16, 1951

Dear Dr. Haldane,

I shall be in England in about the middle of May to the early part
of June, and I hope I will have an opportunity to see you. I would
very much like to meet some of the other members of your department.

I am planning to spend most of the spring and summer in Europe, and
it will start with a visit to Pavia where I will lecture to Buzzati-
Traverso's department. I expect to be there until early in May.

I hope you and Mrs. Haldane are well. With best regards to both
of you

Sincerely yours

E. Mayr (S.P.)

Ernst Mayr

EM:sp

Dr. J. B. S. Haldane
University College
Gower Street
London, W. C. 1   Answered
England

29th May, 1951.

Dr. Ernst Mayr,
c/o Colonel Meinertzhagen,
17, Kensington Park Gardens,
W.11.

Dear Dr. Mayr,

  Professor Haldane has asked me to write and ask
you if you can manage to come here on Thursday as soon after
12 noon as possible, instead of at 12.30 as arranged.

    Yours faithfully,

    Secretary to Prof. Haldane.

THE AMERICAN MUSEUM OF NATURAL HISTORY
CENTRAL PARK WEST AT 79TH STREET
NEW YORK 24, N. Y.

September 26, 1951

Dr. J. B. S. Haldane
University College
Gower Street
London, W. C. 1, England

Dear Haldane,

I am back in New York and trying to catch up with all the things
that were neglected during the past five months. I had a most
profitable time not only in England but also in various other
European countries, and I hope to derive benefit from the many
stimulating conversations for years to come. Next time I shall
certainly plan to stay in England longer.

One of the men I met in you lab, Freeman, works on fleas, if I
remember correctly. Ask him if he has seen a recent paper by
Peus on the geographic variation of Ctenophthalmus argytes,
published in the Kleinschmidt Festschrift (1950, pp. 286-318).
There are quite a few interesting aspects to the geographic
variation of this species as emphasized by Jordan many years
ago. It would seem well worthwhile if Freeman would devote
some attention to this species and, in particular, try to
collect along the borderline in France where the celticus group
and the euros group approach each other. They are supposed to
be completely allopatric and still show no sign of inter-
gradation in the zone of contact. Presumably, they are two
populations that had become isolated during the Pleistocene,
and their present borderline is the line where they met
after their post-Pleistocene expansion.

With kindest regards, also to Mrs. Haldane

Sincerely yours

Ernst Mayr

Ernst Mayr

EM:sp

Return letter to me

UNIVERSITY COLLEGE LONDON

DEPARTMENT OF BIOMETRY

EUSton 4400

PROFESSOR J. B. S. HALDANE

GOWER STREET, W.C.I

3rd October, 1951.

Dr. Ernst Mayr,
American Museum of Natural History,
Central Park West at 79th Street,
New York, 24. U.S.A.

Dear Mayr,

Thank you for your letter. Freeman says C. agyrtes
in England is pretty uniform, and he is not prepared to do
extensive trapping of small rodents in a foreign country,
especially as Rothschild and Jordan, when doing so in France,
were arrested for killing game without a licence!

Meanwhile, could you be so good as to let me have a
reprint of Delacour and Mayr on ducks and geese (Wilson
Bulletin, 1946)? Scott has got a number of interspecific
and intersubspecific hybrids, and my wife hopes to get some
of them suitably mated. Several intergeneric hybrids are also
there, e.g. Anas x Aix, but they are more likely to be sterile.

We greatly enjoyed your visit. We have now rather less
newts than when you were there, as the back-crosses and F$_2$s
have been dying as predicted. We should also be very glad to
have a recent general paper of yours on bird classification of
which I have heard.

Yours sincerely,

J. B. S. Haldane

Large be. paper
my by Yamashina

Send 219, 220, 221, 223
224, also 186

October 23, 1951

Professor J. B. S. Haldane
Department of Biometry
University College
Gower Street, W. C. 1
London, England

Dear Haldane,

The 300 reprints Delacour and I had of the duck paper were, unfortunately, exhausted within one year's time. Obviously ducks are a
very popular group. Peter Scott, in one of his annual reports,
has reprinted most of the classification, and this may be sufficient for your cytological purposes. Yamashina in Japan has
completed in manuscript a long paper on the cytology of ducks.
It is an excellent paper so far as the descriptive aspects are
concerned but was very weak in its general conclusions. Several
of us suggested that he completely revise that general part.
This was more than a year ago, and it appears as if Yamashina
is reluctant to make these changes. I wonder whether you care to
get in touch with him. His address is:

I have sent by separate mail the reprint on the classification
of birds, as well as some other recent papers. The main point
I wanted to bring out in the classification paper is that we
actually know very little about the relationships of the major
groups of birds. Nearly all efforts during the past fifty
years have been directed toward an analysis of intraspecific
variation.

With kind regards, also to Mrs. Haldane

Yours sincerely

EM:sp   Ernst Mayr

UNIVERSITY COLLEGE LONDON

EUSton 4400   DEPARTMENT OF BIOMETRY

PROFESSOR J. B. S. HALDANE   GOWER STREET, W.C.I

2nd November, 1951.

Dr. Ernst Mayr,
American Museum of Natural History,
Central Park West at 79th Street,
New York 24, U.S.A.

Dear Mayr,

Thank you for the reprints. Perhaps the greatest contribution of a biologist to geology was made by Darwin in the first
edition of the Origin of Species (Chap. IX). He calculated
the time from the late Cretaceous till the present as 300 million
years; about four times too long, but at a later date physicists
were giving 10 million years for the whole age of the earth, about
300 times too short. Darwin withdrew this estimate in the 6th
edition. So the biggest contribution made by a biologist was not
accepted. In spite of your remarks I am inclined to be a "drifter"
on the basis of the Permo-carboniferous tillites, especially of
South Africa. But my opinion is based on hearsay.

Yours sincerely,

Check with my reprint
tell him about symposium

November 15, 1951

Professor J. B. S. Haldane
Department of Biometry
University College
Gower Street, W. C. 1
London, England

Dear Haldane,

Yes, you are right that Darwin was perhaps the biologist who made the
greatest contribution to geology. There is much in the ORIGIN OF
SPECIES that is overlooked by everyone, even though it is as valid
today as it was then.

If you look at my paper carefully you will find that I do not deny
continental drift, all I do is deny that it played a role in the
Tertiary. This has now begun to be admitted even by some of the
foremost proponents of drift. We had a symposium on the subject
last year which is now in press. There is no chance for land
connections as demanded by drift after the Triassic.

                                        Sincerely yours

                                        Ernst Mayr

EM:sp   Ernst Mayr

Please give my best to Helen!

May 4, 1954

Professor J. B. S. Haldane
Biometry Department
University College
Gower Street
London, W. C. 1, England

Dear Haldane:

    I hope to be in London for a few days around
May 26, on my way to the International Ornithological Congress in Basel. I am looking forward to
seeing you and Dr. Spurway. If there is any special
day on which you would like to see me or the opposite, please write me in care of David Lack. I
expect to be in Oxford around the period of May 21
to 24.

                        With best regards,

                        Ernst Mayr

EM:ar

UNIVERSITY COLLEGE LONDON · · GOWER STREET WCI

DEPARTMENT OF BIOMETRY

Professor J. B. S. HALDANE

Telephone EUSTON 7050

4th June, 1956

Dr. E. Mayr,
Museum of Comparative Zoology,
Cambridge 38,
Mass.

Dear Ernst,

Thank you very much for writing to me. Haldane said that a little flattery from certain sources changed my mood so much that this revealed how little self respect I possessed.

We apologise for not answering before. This was because we have discussed each other's drafts so that these hung about and did not get sent off smartly.

It is remarkable that your geneticists don't go in for the analysis of fancy breeds. In England the Darwinian and Batesonian interest in them is not quite dead. The pedigree of causes resulting in the round body, doubling of the anal and caudal fins, and the popping of the eyes of the veil-tail telescope goldfish was published in a popular book. The pituitary is histologically abnormal. It is particularly interesting how fancy animals manage to be homozygous for sublethals termed by "modifiers" like the crested fowls, with their more or less cured cerebral hernia, or heterozygous for untamable lethals like the crested ducks and canaries. Both these things happen in speciation, but the latter so far only in plants.

However it can be compared with chromosomal sex determining mechanisms, as the technical term "consort" for the homozygous form implies. The genes which have been collected by selection for the "consort type", which must not only modify the heterozygote for the lethal, but satisfy the fancy when themselves homozygous, might throw some light on the elusive sex-genes to which we impute similar functions.

One of the many distressing things we noticed when beginning to read animal behaviour literature was the common use of "spurious analogy" as if it were part of a smear technique. All analogies are spurious, but the various animal psychologists have never thought of investigating the limitations of these, let alone quantifying them, as an aeronautical engineer quantifies the discrepancies between flying conditions and his extremely useful analogy to these known as the wind tunnel. In biology there seem to be legitimate (i.e. those used by me) and illegitimate comparisons (i.e. those used by you) and words of value judgment, of which the greatest is "homologous". J. Z. Young waits for more and more

-2-

analogies. He points out that all language is analogy and the contemporary respectable analogy for nervous functions is so well imbedded that its analogous nature has been forgotten and it therefore seems to have a moral superidity over newer metaphors even though it is inadequate for the discussion of higher nervous functions. These are still discussed by a still older analogy (i.e. animism) which we have relearnt to criticise. Young points out that we only distrust the new analogies because they literally do not describe very much - we invent a new analogy for every particular case.

On the other hand we (the Haldanes) are annoying the European ethologists by taking their psychohydraulic model seriously, both trying to look for them in the C.N.S. and consistently extending it to more and more aspects of the phenomena - this is a good game if nothing else. Hells-bells it is an honest metaphor, not like Tinbergen's hierarchical diagrams which are a bad mannered misuse of the conventional symbols which should imply the results of the observation of nerve tract degeneration after lesions.'

We hope your pupil will make his ethology a bit quantitative. Our model for quantitative ethology is Moreau (Proc.Zool.Soc.107A or Brit. Birds 39). He could get a hell of a lot more out of his figures with a little statistics. But it is amusing that a bird's times of absence from the nest are more variable than its times of sitting on eggs (as they ought to be - perhaps?). We have a monstrous paper on such variation cooking up on our 1954 climbing perch work which must be ready for publication in Sankhya when we go back to Calcutta in the beginning of July.

Are you by any chance going to the Japanese genetical symposium in September? If not we will look forward to seeing you in 1957.

Yours,

H. Spurway.
J B.S. Haldane

Please remember us to Gretel.

INDIAN ~~STATISTICAL~~

Telegram: STATISTICA, CALCUTTA-35
Telephones 56-3222-27 (6 lines)

No. Bio/H/0,822

203, BARRACKPORE TRUNK ROAD
CALCUTTA-35

March 28, 1959.

Dear Mayr,

    I enclose a sheet showing the visits of 10 insect species to the flowers of six plants of <u>Lantana camara</u> on whose genetics the author, Mr. K.R.Dronamraju, is working. Three (Orange) have flowers which open as yellow and turn orange. Three (Pink) have flowers which open white and turn pink. The genetics are not yet known. Four insect species are not yet identified with certainty. I think it is clear that but for <u>Precis almana</u> the Pink/s would be rather poorly visited. It is also clear that there is no sexual isolation of the two forms, and would not be, even if <u>Precis almana</u> were the only species available. The data for the swallow-tail are interesting. It tended to visit each from several times in succession. The bee of Mar. 22 was not "blumentreu ", that of Mar. 26 was so. This was probably a matter of "learning" not "instinct". There are lots more data scattered round the year.

    Now I want your advice. In the first place the idea is Mr. Dronamraju's. I put him onto the genetics. He noticed the selective visits of the first two species. I then said "This is much more exciting than one more character which does or does not mendelise". I have been quite unable to find anything like it in the literature, though the observations are very simple, and Sprengel might have done it. By the way I doubt if there is a copy of his book in India. Darwin and Lubbock didn't, nor did Dora Ilse. Have you got strong views on publication, or suggestions? He proposes to try a lot of the obvious things, e.g. coloured paper "flowers", and experiments on ~~moths~~ immediately on emergence. Do you think "Evolution" would accept a paper on the subject ? The data are mostly rather less systematic than those given here, but they support them.

    You will see that my time here is not being wasted. I have another junior colleague who has got even more striking results, but not so much in your field. In each case if they had been working in the average university here the supervision would probably have told then to get on with their assignment, and would not have seen that they had got onto something much more interesting. Biology in India is in rather a dull stage at present, but it would be when some young men have grown up, provided that they remember they were young.

    Please then let me know if there is any parallel work which Dronamraju ought to know of, and any comments you may care to make. My wife also sends her kind regards.

Prof. Ernst Mayr.
Dept. of Zoology,
Harvard University,
Cambridge,
Mass.
U.S.A.

Yours sincerely,

J.B.S. Haldane
(J.B.S.Haldane)

x only one species of moth, a "dirty white" is concerned

written to H. S,
Nov. 11

April 13, 1959

Dear Haldane:

I have been planning to write you for the longest time
and I am glad that you have shaken me out of my inertia.
Perhaps this isn't quite the right word because no one seems
to give us poor evolutionists a chance to be lazy and inert
in this year 1959. Travelling from one evolution conference
to the next, I feel like the old-time Vaudeville performer
who travelled from convention to convention and from county
fair to county fair. At least he had the advantage of showing
the same old tricks to ever-new audiences, while I am supposed to say something new each time because every word I
utter is going to be published. I shall celebrate the passing
of this trying year by going to Australia and if all goes well,
I hope to pass through India on my way home. This will not
be until March or April 1960, and there will be plenty of
time to discuss my forthcoming visit. I have long wanted to
come to India where I have quite good friends and I hope this
will finally be possible. Not the least reason will be to see
you and Helen again. The grapevine has it that you both are
enjoying India and that you are well and prospering. It will
be a pleasure to be able to confirm this in person.

And now to the subject of your recent letter. It seems
to me that there is no question about the facts. A definite
reference is clearly evident. We must make a distinction between the preference of a social bee which may change from
day to day and that of most solitary insects which, in many
cases appears to be part of the code of information handed
down by previous generations. I gather that most of the insects in the Lantana case are of the latter sort. I know
of no similar case in the literature but have written to
one or two people to find out whether they know some cases.
The closest I rmemeber is the case published by Verne Grant
where insects showed a definite preference when two subspecies
of a species of Gilia were randomized in a flower bed. I
have asked Grant to send you a reprint of this paper which
was published in one of the earlier volumes of Evolution.
Mather in his work on Antirrhinum dealt, I believe, only
with social bees. I asked one of von Frisch's students,
Dr. Lindauer, and he has no information on the subject.
I really think that it will be worth while for you student
to follow this up even though there may be something about
it in the earlier literature. The worst that can happen is

that these forgotten facts be rediscovered. There is, of
course, good evidence that bees have species specific preferences for certain flowers. It is also known that in certain insects, particularly in crepuscular species, there is
a far greater sensitivity to ultra-violet than there is in
man. It might be worth while if your student were to photo-
graph these two flowers of <u>Lantana</u> through filters that screen
out all the visible part of the spectrum. Perhaps one of
the two kinds of flowers is much richer in the ultra-violet
range of the spectrum. I can see all sorts of experiments
that could be done. In short, I entirely agree with you
that this is a most interesting field and that your student
should be encouraged by all means to follow it up. I am
sure that Evolution would be interested in a paper on the
subject. You know of course that Michael Lerner is now the
Editor.

      I would greatly appreciate it if you would place me on
your mailing list. I have not had any reprints from you in
quite some time. Indeed, there are one or two of your recent
papers which I have not yet been able to read because every
time I went to the library the particular issue of the Journal
of Genetics was lent out to a student.

      I recently paid a flying visit to Gower Street where I
visited Maynard Smith and discussed with him his selection
experiments. It is rather interesting to see how the interaction of genes becomes the central problem wherever one
checks what they do in some genetic laboratory. I must say
that University College didn't seem to be the same place
without you and Helen.

      I have been asked to write so many general papers that
it becomes high time for me to get back to "honest zoology".
I have a lot of things I would like to do as soon as I get
my head over the water. Three of my students work on behavior
problems but all of them are field naturalists by inclination
are so fascinated with recording previously unrecorded behavior patterns that I am afraid I will have to let them go
through this stage of their education. Indeed, they learn
observing which is perhaps the most important scientific technique,(the claims of some experimentalists notwithstanding).
Also they do record their data quantitatively and at the worst
will produce the material with which someone else will construct
fancy theories.

      We spent the last weekend at our retreat in New Hampshire
where a beaver colony established itself within 150 yards of
our house. This is where I would like to spend my time rather
than down here in Cambridge. Too bad one has official duties!

      With the best wishes to both of you,

                              Yours,

                        Ernst Mayr

Wr. Yec. 21

INDIAN STATISTICAL INSTITUTE

Telegram: STATISTICA, CALCUTTA-35
Telephone 56-8222-27 (6 lines)

No. Bio /H/ 25 11

208, BARRACKPORE TRUNK ROAD
CALCUTTA-35

18 November 1959.

Dear Mayr,

 I am writing to you as Helen has broken her scaphoid. When coming here from Singapore:-

 1. Give us plenty of notice, if you would like to go to Darjeeling, to the Sanderbands (Deltaic islands) or anywhere else where you may see an ecology more "natural" than that of the neighbourhood of Calcutta, though we are not badly off for birds here. Also I hope you will spend a day or two here looking at Dronamraju's work on selective visits of insects to flowers. The following are records of numbers of visits of 7 species to "Orange" and "Pink" plants of <u>Lantana camara</u>, spread over many hours during many months.

Species	Orange	Pink
Precis almana	218	13
Danais chrysippus	142	152
Papilio polytes	15	31
Papilio demoleus	42	98
Catopsilia pyranthe	40	603
Baoris mathias?	1	108
Apis indica	71	245

 Besides these a number of species have made under 40 visits.* The conditions varied, but there were about equal numbers of pink and orange flowers. Helen and he are working on P.demoleus of which larvae are common, immediately on emergence. They show a rather stronger preference for pink than appears from this table. The bee results mean nothing. A bee kept to one colour on a given day. The Lepidoptera were more random in their visits.

 He is writing his results up for "Evolution". But I am sure that a discussion with you will be most valuable.

 2. Do not book by B.O.A.C. or Quantas. They are constantly breaking down, and have no planes to replace those which are taking a rest. This means that one cannot book air or rail tickets in advance at this end.

 The weather is just getting nicely warm about March 1st. Give our kind regards to White and his wife and your wife. Is she with you, we hope so?

* Two which seem to favour Orange, are Nymphalids like P. almana.

Prof. Ernst Mayr
C/o Prof.M.J.D.White,
Dept. of Zoology,

Yours sincerely,

J. B. S. Haldane
(J.B.S.Haldane)

INDIAN STATISTICAL INSTITUTE

Telegram: STATISTICA, CALCUTTA-85
Telephones 56-9222-27 (6 lines)

No. B1./H/2629

203, BARRACKPORE TRUNK ROAD
CALCUTTA-85

Dear Mayr,   30 December 1959

We are delighted to hear that you are coming in February. It is the nicest
month of the year in Calcutta, though beginning to warm up in Orissa. A draft program
for Orissa is:-

2 hours flight from Calcutta to Bhubaneshwar. See (a) large group of Hindu
temples (no admission to largest, but lots to see elsewhere), (b) a group of Jain caves
and a modern Jain temple.

Car to Konorak. Monstrous temple of Sun, completely covered with sculpture.
This is largely "erotic", but I should call it acrobatic. Car thence to Puri.

Puri. Bathe. See several temples from outside and possibly car of Jagannath .
Buy souvenirs (sculpture, textiles, playing cards for a very queer game.)

Return Bhubaneshwar.

We can probably do it all in two days. If, as I hope, you want to do some
bird-watching, there are at Bhubaneshwar (1) Low hills with deciduous trees, (2) Dry
shrubby country, (3) permanent water near the temples. Puri is a good deal moister,
and Konorak intermediate. Dronamraju could accompany us. Of course however if Biswas
has been asked to arrange this trip for you, I don't want to interfere with his
arrangements.

An important point is this. Try not to come by BOAC and QANTAS. They are very
unreliable. The last man to do so was late, and we had to cancel an air trip to Assam.
Also try not to arrange thing so that you have to be in Orissa on a Wednesday or
Thursday, if you are going with us, since Helen will have a practical class on one of
those days.

Helen and I hope that you will stay with us. We have a flat about half a mile
from here. As we only moved this month the spare bedroom is still a dump. But it will
be ready/ by February. I warn you however that you will get Indian food (vegetarian
or not as you prefer, and if vegetarian, Bengali or South Indian) and no bath, but a
fairly efficient shower. So if this is too alarming I can get you invited as a guest
of the Institute, where you will get imitation "Western" food, air conditioning (perhaps)
a radio, and so on. In return for which they will expect you to lecture. There are no
strings on our invitation.

My hope is that you will be able to go back by Bombay, and meet our great
ornithologist, Salim Ali. I suspect, too, that the Bombay Natural History Society's
collections are in some respects superior to those of the Zoological Survey in Calcutta.
He is just starting on the first serious work on bird migration to be done here.

You see we have left you lots of choices. Of course if you like you can stay
in Calcutta hotel. Part of one fell down recently. But they have a wider choice of
drinks, whereas we only have five kinds, all made in India, and none as bad as much
of the imported whisky. But while the hotel is nearer Biswas, it is a long way from us.

Holl.

-2-

So whatever happens you will need cars and/or taxicabs. So please stay here, if only
because we are open till midnight, and the best time for bird-watching and bird-listening
is early morning. Our garden contains spiders which weave webs whose threads are
arranged in rectangles. I bet they haven't got such a web in the Zoological Survey! So
try to stay with us.

Helen's fracture was almost painless. Since then an elephant has brodden on
her toes, and a tiger patted her (both gently).

Prof. E.Mayr.   Yours sincerely,
C/o. Dr. D.L.Serventy,
C.S.I.R.O. W.Austr. Regional Lab.
University Grounds   J. B.S. Haldane
Nedlands, W.A.   (J.B.S.Haldane)
Australia.

March 8, 1960

Professor J. B. S. Haldane
Indian Statistical Institute
203 Barrackpore Trunk Rd.
Calcutta 35, India

Dear Professor Haldane:

Many things have impressed me during my recent visit to India, some for which I was well prepared such as the wonderful temples of Orissa and some of them quite unexpected. In the latter category I would like to mention the Calcutta Zoo. In my travels all over the world I always make it a principle to visit the zoo in any major city. These zoos generally contain the obligatory large mammals and in addition a few spectacular snakes and water-birds. Among the many zoos that I have visited only very few have as rich a collection of the smaller and more delicate species, particularly of birds, as the Calcutta Zoo. Yet your zoo impressed me not only by the wealth of rarities, but also through the efficiency and competence with which it is run. All the animals I saw seemed to be properly housed and fed according to the best scientific principles.

I have little doubt that the present superintendent, whose name I believe is Lahiri, deserves credit for having given Calcutta a Zoological Garden of which it can be truly proud. I have been privileged to see Mr. Lahiri at work and this has greatly increased my admiration for him. He is unusually observant and discovers at once when something is not quite as it ought to be. Let me quote to you only one example. While our group watched some gibbons on Gibbon Island, I noticed out of the corner of my eye that a swan had difficulty climbing on to the steps leading out of the moat around the island, because the water level was a little low. At once I heard Mr. Lahiri, who apparently had made the same observation, give the order to add water to the moat. The speed with which he not only diagnosed the difficulty but also corrected it was quite remarkable. There were two or three other occasions during my tour through the garden when Mr. Lahiri with equal dispatch and decisiveness corrected minor shortcomings.

I would greatly appreciate it if you could bring my high opinion of the Calcutta Zoo to the attention of the responsible authorities. I would like to congratulate the City of Calcutta on having a Zoological Garden which is equal to the best in the world. I would also like to record my admiration for the efficiency with which Mr. Lahiri carries out the duties of his office, for his unusual understanding of the needs of wild animals, and for the devotion which he is giving to his demanding job.

Very sincerely yours,

Ernst Mayr
Alexander Agassiz Professor of Zoology
at Harvard University

EM:nt

Telephone No.45-4150   (COPY)   ZOOLOGICAL GARDEN,ALIPORE.

Letter No. D.O. 2464   Calcutta, the 25th March,1960.

Dear Professor Haldane,

     I am extremely thankful to you for your kind letter No.Bio/H/2885
dated 15th March,1960, forwarding a copy of letter dated 8.3.60 from Dr.Mayr
of Harvard University. I feel proud indeed to find that our exhibits, their
mode of keeping and management impressed the eminent Ornithologist. The
high appreciation expressed by Dr. Mayr to your goodself will always enthuse
me and my colleague in our work for years to come.

     Permit me to express my sincere thanks to you for the keen interest
you have shown in our Garden. May I request you to kindly convey my regards
and gratitude to Dr. Mayr for his kind letter of appreciation which will
inspire us in our work.

     This letter of appreciation will be placed before the Committee at
their meeting to be held to-morrow.

     With kindest regards.

                        Yours sincerely,
                        Sd/- R. K. Lahiri.

Prof. J.B.S.Haldane,
Indian Statistical Institute,
203, Barrackpore Trunk Road,
CALCUTTA-35.

INDIAN STATISTICAL INSTITUTE

Telegram: **STATISTICA, CALCUTTA-35**   **203, BARRACKPORE TRUNK ROAD**
Telephones: **56-3222(9 lines)**   *No. Bio/4/2939*   **CALCUTTA-35**

28 March 1960

Dear Mayr,

    I enclose a copy of a letter from Lahiri. I
have little doubt that it will help him considerably.
For some nasty criticism has been directed at the garden.
I am trying to collect some reprints to send you. I think
it most important that foreign visitors should record
their appreciation of good work done in India, and quite
equally important that they should not record their appreciation of bad work. If, like myself, they settle down
in India, they perhaps have the right to criticize the
bad work actively. If, for example, I am ever asked to
express my opinions on the Bombay zoo, I shall do so
with some venom.

Prof. Ernst Mayr,   Yours sincerely,
Department of Zoology,
Harvard University,   J. B. S. Haldane
Cambridge, Mass.   (J.B.S.Haldane)
U.S.A.

Encl: 1

Haldane

January 2, 1961

Professor J.B.S. Haldane
Indian Statistical Institute
203 Barrackpore Trunk Road
Calcutta 35, India

Dear Haldane,

Your shipment of reprints reached me on Christmas eve and was my Christmas present that I enjoyed the most. I have been reading and rereading your papers and find them as usual full of stimulating ideas. Independently, I pointed out recently that Bateson's work on bolting in sugar beets was the same phenomenon as Waddington's genetic assimilation.

I am particularly glad to have your paper on the cost of natural selection which I had of course read soon after it had come out. I think it will form the basis of a lot of future investigations. There are two aspects that puzzle me. First of all, what effect does density-dependent mortality have on fitness? If a lot of genotypes survive during a period of low population density, how is the fitness of the genes of such genotypes to be calculated during high densities? Perhaps I should phrase my thoughts a little differently. Population density is highest at the moment of zygote formation. As zygotes die one after the other during development (particularly conspicuous among the lower animals) density-dependent mortality may shift in its impact from one genotype to another. With well over 99% of all the zygotes dying anyhow in many lower organisms (much of the mortality of course being non-selective) deleterious genotypes can be eliminated rather rapidly. A second point that puzzles me is the effect of synergistic vs. antagonistic interactions of genes. In view of the fact that no gene has an absolute selective factor, the contribution of a gene to mortality may be greater when it is combined with one or more other "bad" genes than when it is combined with fitness-increasing genes. As a result a single case of genetic death may simultaneously eliminate several deleterious genes from the population. Wouldn't that change your calculations? I rather suspect that lethal genes are a special category and that what is true for them should not be extrapolated to other genes, at least as far as turnover in the gene pool is concerned.

You may gather from these remarks that I have been greatly stimulated by your papers even though my thinking is not necessarily particularly incisive. At any rate I am most grateful for you sending me these papers. By the way, they reached me by a miracle. Your clerk addressed them to 11 Chancy Street, Massachusetts, USA. I have always maligned our postal clerks for lack of imagination but in this case apparently someone must have figured that a parcel of reprints from India was most likely destined for a Harvard Professor, and so they tried Cambridge.

It is almost a year since my visit to Calcutta. I remember with great pleasure my stimulating visit with you and
last but not least the wonderful excursion to Orissa. I hope
it wasn't my last chance to see India, indeed I would like
to stay longer the next time. Please give my best regards to
Helen and to Dronamraju.

Best wishes for 1961,

Yours sincerely,

EM:ehw   Ernst Mayr

INDIAN STATISTICAL INSTITUTE

No. Bio/H/4027

Telegram : STATISTICA, CALCUTTA-35
Telephones : 56-2222 9 lines)

208, BARRACKPORE TRUNK ROAD
CALCUTTA-35

Dear Mayr,   11 January 1961

  I am still thinking about the "cost of natural selection" or the "substitutional load". I have once trivial paper in press, giving accurate expressions for it. In another, not yet in press, I tackle the question of natural selection of slowly changing intensity (e.g. during a slow climate change). For reasonable values this does not greatly diminish the "cost". Your theorists, such as Kimura & Morton, distinguish 3 genetic "loads" :—

  (1) Mutational

  (2) Segregational (against homozygosis)

  (3) Substitutional (cost of selection)

I add :— (4) Dispersional.   I add :—

  If (in any particular year) I add each microhabitat contained only the gene frequencies (defined by minimizing (2)) best adapted to it, deaths would be postponed. But organisms (including pollen grains) migrate or disperse out of such areas. Plants with relatively heavy seeds, and fairly perennial (e.g. Plantago spp.) can establish different ecotypes a mile apart. As I have pointed out, genes making for resistance to density-dependent factors will tend to drift outwards from crowded areas more than genes for resistance to tough environments will tend to drift inwards.

  (5) Dyszygic (or some such word). The segregational load can sometimes be greatly cut down if all loci with heterotic effect are completely linked (as in 7-ringed Oenothera "species"). The load is the extra mortality due to this not being done. More generally, for any given set of segregating genes in a given environment (or a given set of fitnesses of genotypes) the gene frequencies _and_ the linkage patterns can vary. The optimal frequencies must depend on the linkage patterns. But there is an optimal combination of gene frequencies and linkage pattern. And we lose by not having got it. Linkage patterns are always stable. Selection ? always opposes new translocations and generally new inversions. I have not thought this one out.

  (6) (No name yet). Read S.K.Roy's paper in the last Journal of Genetics. In a self-fertilizing plant population there must be some "optimal" combination of different genotypes. But it is probably unstable. This is probably enormously important in our species. I think it is quite a good thing to have a few people like, say, Wagner or Gauguin, But even 1% would perhaps be too many.

  I may be in this class. I have just, as you know, been awarded the National Academy's Kimber Medal. I should like to come in April and collect it. But when I last applied for a visa to your country I was told to list all organizations of which I had been a member since the age of 16, with dates of entering and leaving them. I regret that I do not know in what year I joined the National Mouse and Rat Club, to mention only one. And I think a scientist should go thru the motions of telling the truth. So unless your new State Department takes action, I fear I can't come. But if everybody had scruples of this kind, civilized life would be impossible.

  I will think over your other remarks. Here are some ideas to be going on with.

Prof. Ernst Mayr,   Yours sincerely,
Museum of Comparative Zoology,
Harvard University,   P.T.O.   J. B. S. Haldane
Cambridge 38, Mass. U.S.A.   (J.B.S.Haldane)

February 21, 1961

Professor J.B.S. Haldane
Indian Statistical Institute
203, Barrackpore Trunk Road
Calcutta-35, India

Dear Haldane,

Thank you for your letter of January 12. Yes,
there are numerous genetic loads. I know of several
others in addition to those mentioned in your letter,
not counting the various kinds of genetic loads that
occur more commonly in plants than in animals. I have
prepared a first draft of a discussion on genetic load
for my forthcoming book and am enclosing it for criticism. Perhaps this will make some of the questions in
my last letter a little clearer. The relativity of
fitness and the possible synergistical action of favorable as well as deleterious genes is not sufficiently
coped with in published calculations. Our interest in
questions of this sort is so recent that we still lack
the pertinent facts. It is quite amazing how little
we know about the genetic variability of wild populations, except for a few species of Drosophila and a
few polymorph loci in some other species. I think this
is where a lot of research will have to be done.

Please thank Krishna for his report and letter.
I will write him presently.

I hope you can come in April, without too much
annoyance from our bureaucracy. Having done a lot of
travelling in my life, I know that there is hardly a
country that does not have a ridiculous bureaucracy and
no one is more embarrassed by their antics than the decent citizens of that country. I rather suspect that our
current administration will try its best to be sensible
in such matters.

With best regards, also to Helen.

Yours sincerely,

Ernst Mayr

EM:ehw

modified
f/d c

Haldane 3/10/61 ⑥

INDIAN STATISTICAL INSTITUTE

Telegram : STATISTICA, CALCUTTA-35
Telephones : 56-3223 (9 lines)

No. Bio/H/4266

203, BARRACKPORE TRUNK ROAD
CALCUTTA-35

3 March 1961

Dear Mayr,

Thanks for your letter and enclosure. It does not, of course, matter, but I think I may fairly claim priority for specifying loads (1) and (2) and estimating (1). In the American Naturalist (1937), LXXI, 337-349 I estimated the mutation load as about 4% for <u>Drosophila melanogaster</u> and 10% for man. I think both figures are too low. I am sorry I have no reprints left.

If I were you I would perhaps make the point that a "load" even of 50% may be slight if it occurs early enough in life. You nearly make it on p.3. but not quite. A death of half the zygotes before they have begun to eat imposes no load on the species food supply beyond that needed to make half the eggs. If, say, lethals killed the same number during pupation, this would be much "heavier". The inviable eggs may even be eaten by their brothers and sisters, and inviable babies by their mothers.

Dronamraju has just gone to take up a job in Glasgow. He has a paper with Meera Khan in press in the Journal of Heredity. But they now have much better figures based on hospital patients who may be expected to be rather inbred, and school children, who may be expected to be rather outbred, 7.3% of all marriages are with nieces, 16.8% with first cousins. I think, if we can get support for this work, we may be able to estimate some loads. But it is less trouble to do a job of work than to cadge for the cash.

I haven't thought much about loads lately, as some broken bones have been troubling me.

Professor Ernst Mayr,
Museum of Comparative Zoology,
Harvard College
Cambridge 38, Massachusetts,
U.S.A.

Yours sincerely,

J. B. S. Haldane
(J.B.S.Haldane)

March 16, 1961

Prof. J.B.S. Haldane
Indian Statistical Institute
203, Barrackpore Trunk Road
Calcutta-35, India

Dear Haldane,

Thank you for your letter of March 3. Yes, the difference between early and late mortality is precisely the one I was trying to make in my own comments, and this is why I think that the "cost of evolution" is less severe than you suggest. If you make the further assumption, which I think is not altogether unreasonable, that various fitness-reducing factors are synergistic, you may have quite an elimination of genes and genotypes without any real threat to the survival of the population. It is only in a few species with parental care that we have the very low production of zygotes as is typical for man. On the other hand, it is in these species that accidental mortality is largely eliminated, which is perhaps the explanation for the rapid evolution in such types as the elephants and hominids with their low reproductive potential.

My main point really is that one must eventually go beyond the preliminary stage of assuming that genes have constant and absolute selective values, an assumption we all know not to be realistic but into which we are being forced by the simplifying value of the assumption. My feeling is that by operating with "average selective values" of genes, we introduce quite unrealistic models into our calculations. The mere fact that alot of the early mortality of zygotes is relatively unimportant for the maintenance of populations is one of the reasons for my doubts.

I think I mentioned in my last letter that the Kennedy Administration was far more reasonable in its international policies than the preceding Administration. I just saw in our newspaper that a complete change of the passport application system has been introduced, which, I am sure, will make it quite simple for you to get a visa to enable you to receive the Kimber gold medal in person. I shall attend the award ceremonies and am very much looking forward to seeing Helen and you on your visit to the States.

With best personal regards to both of you.

Yours sincerely,

Haldane

INDIAN STATISTICAL INSTITUTE

Telegram : STATISTICA, CALCUTTA-35
Telephones : 56-2222 (9 lines)   *No. Bio/H/4121.*   203, BARRACKPORE TRUNK ROAD
                                                                 CALCUTTA-35

                                                                 19 April 1961

Dear Mayr,

I am very sorry I shan't be able to get to Washington on April 24th. But my left foot has certainly not recovered from two fractures. Roughly speaking, to go there first class, which, given my foot, I need, with my wife as escort, would cost me a year's income. And between the Indian and American bureaucracies I doubt if I could have got a visa. The Indian one has taken nearly a year getting me Indian nationality.

I have sent Dryden a copy of the speech which I should like to have made. He or someone else may suppress it as not being sufficiently pompous. I send you a copy so you can leak it out in this case.

The clue to paragraph 1 is that I really admire W.E.Castle. In the great days of classical genetics one could be sure that Castle's results would be repeatable. One could also bet four to one that if Castle backed one of two possible theories, the other one would be right. But he piled up facts about rabbits and rats. And they needed piling. Punnett is a similar person in England. One is apt to forget now how difficult their work was when they did it.

Apart from my foot, Helen and I are very well. It is quite hot, and the power supply here is constantly breaking down, so the fans stop.

Dr. Ernst Mayr,   Yours sincerely,
Museum of Comparative Zoology
Harvard University   J. B. S. Haldane
Cambridge 38, Mass., U.S.A.   (J.B.S.Haldane)

April 26, 1961

Dr. J.B.S. Haldane
Indian Statistical Institute
203, Barrackpore Trunk Road
Calcutta 35, India

Dear Haldane,

We missed you on the occasion of the award
ceremonies. However, under the circumstances, you
probably had no choice but to abandon your original
plan. Five other medals were awarded on the same
evening and, as is customary with our National Academy,
none of the recipients responded with a statement.
However I shall take the liberty of sending copies to
some of your friends. I am sure you will like them
to know what you would have said if there had been an
opportunity and at the same time they will appreciate
hearing from you in this manner.

I am particularly sorry to learn that your
foot is not yet entirely back in good working order. I
hope that it will not interfere too much with your work
particularly your field work.

With best regards, also to Helen.

Yours sincerely,

Ernst Mayr

EM:ehw

NAS

April 27, 1961

When J.B.S. Haldane learned that he was to be awarded the
Kimber Genetics Medal at the 1961 meeting of the National Academy
of Sciences, he jotted down a few typically Haldanian remarks
"suitable for the occasion." As it turned out his foot injury made
the trip to the United States impossible, not to mention the fact
that it is not customary for the recipient of such medals to make a
response.

As a result these comments were "still born." Having a copy
of the response, I thought I would make it available to those who
might be interested in this specimen of Haldania.

(Ernst Mayr)

"In thanking the Academy for the honor bestowed on me, I
should first like to say that, although I have been awarded several
other medals in the past, I have always felt that I would have been
still prouder if one or two names had been omitted from the list of
my predecessors. For the first time in my experience this is not so
today. I am proud to be classed with each one of the previous recipients of the Kimber Medal.

I am told that someone is to speak of my alleged scientific
achievements. The list of them given will perhaps be more objective
than any which I can give. But I should like to mention three which
please me because of their simplicity. The first is my demonstra-
tion that moths, rats, and cress plants contain a substance now
called cytochrome oxidase and first discovered by Warburg in yeast,
for which carbon monoxide competes with oxygen. The second is my
discovery, with E.M. Case, that oxygen, so far from being a colorless,
inodorous, and tasteless gas, has quite a taste at six or seven at-
mospheres. I paid for this with two fractures of the vertebral
column. The third is the calculation of the first twelve cumulants
of the binomial probability distribution, and the recurrence relation
between them.

Had I not made them, most of my discoveries would have been
made by someone else within a few years. Thus W.E. Castle discovered
linkage in rats the year after I did so in mice; and L.S. Penrose
and I measured human linkage simultaneously. But I take full credit
for the discovery both of Subodh Kumar Roy and of Krishna Rao Dronamraju. Neither of these Indians has obtained first class honors or
a doctorate. So they would have had no future in India and no possibility of study abroad but for my intervention. S.K. Roy has discovered that the morphology of some plants at least becomes more
variable at the end of each flowering season, and that the planting
of mixtures of some rice varieties leads to an increased yield of
26% or so. Other mixtures yield less than the mean of the two pure
lines. The genetical analysis will probably be more complicated,
and certainly slower, than that of the "killer" character in Para-
mecium. K.R. Dronamraju discovered that different species of pol-
linating insects have their own preferences between color varieties
of the same plant species; with my wife he found that these preferences are inborn: the analogy with Darwinian sexual selection is

2.

obvious. He has also got pedigrees which make the location of a gene
on the human Y chromosome highly probable, and, with Merra Khan,
discovered the most inbred large human population, ten million or so
people among whom about 7 percent of all marriages are with sisters'
daughters, and 16 percent with first cousins.

If I may be allowed some stop press news, S.D. Jayakar and
I have just shown that the expectation of any of Fisher's "k statistics", or cumulant estimates, in a sample, is equal to the corresponding statistic in the population from which it is drawn; so the
theory of sampling from finite populations is greatly simplified.

I hope that it may be possible for me to spend most of the
dollars accompanying the medal on scientific research in India. For
various historical reasons it is extremely hard to obtain funds for
this purpose without restrictions which paralyse research more or
less completely. Yet perhaps nowhere in the world are there more
plants and animals which seem to shout questions at one, or more
brilliant young men and women whose capacities for research are not
used.

In conclusion, I once more thank the Academy most heartily
for the honor done to me.

                    (J.B.S. Haldane)

-8-

Haldane

COUNCIL OF SCIENTIFIC AND INDUSTRIAL RESEARCH
GENETICS AND BIOMETRY RESEARCH UNIT

Barrackpore House, Barrackpore,
West Bengal.

No: H/252   February 19, 1962

Dear Mayr,

As you see from the address I show some signs of getting into a place of which I am boss. As soon as we get a fence round it, we shall buy some <u>Gallus sonneratii</u>, and I will also take steps to get <u>G.lafayettii</u>, <u>G.varius</u>, and <u>Anas poecilorhyphes</u>, with which to make some fairly obvious crosses.

The most exciting stuff is however likely to be plants for the first year or two. All sorts of allegedly actinomorphic and zygomorphic flowers turn out to be a bit asymmetric when you look at them carefully. But the first person to do so is one of my Indian colleagues. Helen, for the first time, has an almost complete record of the building of a nest by a solitary wasp, <u>Scelephron madraspatanum</u>. About 940 visits were classified and timed, as were the periods of absence before them. On the whole it does take longer to do what seems to us a skilled job, such as making a lid, than an unskilled one such as daubing. It also takes a little longer fetching the mud. Of course she is able to contradict all earlier workers. For example this beast caught a lot of spiders, but also 3 flies. And sometimes it stopped off from job A to do a little on B, and then went back, and so on.

The object of this letter is as follows. As you know I got some cash from the Nat.Ac.Sci, and $500 are left in Washington. I have been asked to bank them. If so I shall have (1) to get leave from the Reserve Bank of India, and (2) to report the State of the account twice annually. I want it for subscriptions to Genetics, Evolution, etc., perhaps even the American Naturalist. Can you take charge of it for me, and pay the relevant subscriptions? If so I will tell the guy at Washington to send it to you.

We are in the throes of a general election here. Even now at 11 p.m. people are parading the streets yelling (or perhaps singing) political slogans. However nobody has been killed yet, though a few noses have been broken. This is not too bad, as we have $> 2 \times 10^8$ voters. I suppose before it is over someone will have his head bashed in. But seeing we have 14 major languages, and 5 major religions, besides Buddhists, Parsees, animists, Marxists, and what not, we really behave ourselves rather well. We are however rather annoyed by certain Americans, such as Keeland, Mohr, and Clarridge, who are intervening in our elections. That kind of thing may be efficient in Siam, but is not likely

P.T.O.

-2-

to get them very far in a country with our size and
traditions.

Salim Ali is at last ringing birds in a big
way. He has had one wagtail (<u>Motacilla</u> sp.) picked
up near Moscow, and one near Irkutsk. He is now interested
in a route through the Brahmaputra gorge, which a lot
of small birds seen to take.

Please let me know about the cash. My wife
sends her kind regards.

Yours sincerely,

J. B. S. Haldane
(J.B.S.Haldane)

Prof. Ernst Mayr
Dept of Zoology
Harvard University
Mass.
U.S.A.

March 1, 1962

Professor J.B.S. Haldane
Genetics and Biometry Research Unit
C.S.I.R.
Barrackpore House
Barrackpore, West Bengal, India

Dear Haldane,

I am delighted to see that you are well established at your new little empire (or shouldn't I use such a dirty word).

It will be a pleasure to take care of your cash and to pay from it dues for scientific journals, etc. If you drop the respective treasurers a note, they can send the dues notices to me directly.

I am glad that the news from Helen and you is so cheerful. There is such a pleasure in doing straightforward research. I am most anxious to get back into it after working for so many years on a book manuscript. The book is now just about finished and has to be delivered to Harvard University Press on April 15. You won't find much in it that is new to you, but I have made a real effort to sort and present systematically a vast amount of scattered information and theory about species and about evolutionary phenomena on the species level. I hope it will be considered for what it is, a progress report, and that it will stimulate people to continue where I leave off. I am always afraid that people might consider as final what is one's provisional temporary conclusion.

I am very much looking forward to the analysis of Helen's observations on <u>Scelephron</u>. I rather suspect that the best workers on hymenopteran behavior have always realized that there was more variability than appears from the textbooks. I have quoted some references to that effect in my paper for the Roe and Simpson volume on Evolution and Behavior. If behavior was a rigid as sometimes claimed, how could natural selection have any material available for the evolutionary changes which unquestionably take place all the time?

With best wishes, also to Helen,

Yours,

Haldane

November 19, 1962

Professor J. B. S. Haldane
Genetics and Biometry Laboratory Unit
Government of Orissa
Bhubaneswar - 3
Orissa, India

Dear Haldane, dear Helen,

I had not heard about your moving to Bhubaneswar until
I received your reprints. I would think that this is a much nicer
place to work than the outskirts of Calcutta. I still remember
with great pleasure my visit to Bhubaneswar under your guidance.
I remember the mornings of bird watching out in the fields, seeing
my first   and studying the activities of the termites in
destroying leaf mould and vegetation debris. I do hope you find
happiness in these auspicious surroundings.

I have just completed reading galley proof of my _Animal
Species and Evolution_ and I hope that it will be available by
about April next spring. I am sure you will find some things in
it that will interest you and others with which you will disagree.
I have been trying to stress the unity of local gene pools and this
entails a certain amount of tightrope walking. I wanted to get
away from the atomistic treatment of populations as samples of gene
frequencies (independent of each other, so to speak), and I had to
be careful not to say anything that would suggest that I was in
sympathy with holistic or finalistic ideas which, of course, I am
not. Well, you will see in due time.

This must be a trying time for India. Regardless of the
facts of the situation, which I cannot evaluate from this distance,
it seems a pity that China decided on such one-sided action.
Curious how all big nations sooner or later become imperialistic.
I presume the Chinese will in due time claim all of Siberia as
part of their Asiatic domain, taken from the Asiatic country as a
result of Russian colonialism.

Tell Jayakar that I enjoyed his Cardium paper which shows
how easily interesting situations can be analyzed, abundant material
being right in front of our noses.

With best wishes,

Yours,

Ernst Mayr

EM:ehw

H

February 25, 1963

Professor J. B. S. Haldane
Genetics and Biometry Laboratory
Government of Orissa
Bhubaneswar - 3
Orissa, India

Dear Haldane,

        I have managed to induce Harvard University Press
to send you a set of page proof of my Animal Species and
Evolution. They promised to send it by air mail and thus
you will have a review copy long before anyone in England
or continental Europe. I have also asked them to add proof
of the bibliography, which should be rather useful to your
students. As you will notice, I have tried to counteract
the modern trend to forget all about the pioneers. In a
number of areas I have made a real effort to trace backward
the development of a concept to its earliest beginnings.

        I hope you and Helen will enjoy reading the volume
and will not find too many things in it that you will have
to disagree with.

                        With best regards,

                        Ernst Mayr

EM:ehw

GENETICS AND BIOMETRY LABORATORY
GOVERNMENT OF ORISSA

No. 0259

BHUBANESWAR-2
ORISSA, INDIA

April 6, 1963

Dear Mayr,

Thank you for the proofs, which I am reading. Whether people agree with your conclusions or not, your book will be an invaluable guide to the literature. As a "bean-bag geneticist" I think your view of a species may be a little too physiological and not historical enough. No doubt the various genes (as regards frequency and location as well as molecular pattern) have to fit together to form, if not an adaptive peak in Wright's sense, a range of such peaks, separated at most most times from other such ranges by deep valleys. But on reading you I sometimes get the feeling that you think we could calculate the species if we knew enough about the genes. My guess is that existing species are only a small fraction of those which might have been made up with the genes available in a genus or family, and that the reasons why we have just these species are largely historical.

As will be seen from p.400 of the enclosed, I think your definition of a species, though not your description of it, is a little too futuristic for my taste. I tried to develop the philosophical side of my view of species in an article called "Differences", (Mind LVII N.S. July 1948, pp.294-301) but I have no offprints left.

A pair of _Ianellus_ (or _Lobipluvia_) _malabaricus_ is now nesting in my garden and can be seen from the veranda with field glasses. S.D. Jayakar and H. Spurway are watching them for 13 hours daily, noting down all journeys. At present they seem to be mainly occupied in keeping 4 eggs cool, both by shading them and by shading them and by welting their feathers and hence the eggs. This job will go on for another month or more unless one of the numerous predators such as Herpestes, Varanus, and Canis succeeds in terminating it. However as another nest not far off has produced some chicks, these birds have a fair chance. I know of no comparable observations on ground-nesters. We are within about 1 km of the area where you saw larks and other birds when you were here. Perhaps you will come again. But you might find us unhospitalble during daylight.

On looking at the reprint, I find it contains a nice Greek quotation for anyone who wants to diminish the sales of a book on bird taxonomy or migration.

Prof. Ernst Mayr,
Museum of Comparative Zoology,
at Harvard College,
Cambridge 38,
Mass.,
U.S.A.

Yours sincerely,

J.B.S. Haldane
(J.B.S. Haldane)

File   Haldane

GENETICS AND BIOMETRY LABORATORY
GOVERNMENT OF ORISSA

BHUBANESWAR-3
ORISSA, INDIA

No. 0337

May 8, 1963

Dear Mayr,

Thank you for your reprints. To show that I have read
them, let me do some proof correction. "David" in the
bibliography of "The Emergence of Evolutionary Novelties"
should be Davis. Similarly "Jenkins" should be Jenkin.
Our local weaverbirds, Ploceus philippinus put some mud in
their nests, allegedly to weight them. They are said to
stick fireflies in the mud for illumination. I am waiting
to see one.

I am an unrepentant "beanbag geneticist". Non-mathe-
maticians often fail to realise the complexity of behavior,
and sometimes the self-regulatory capacity, of material
systems composed of simple components. Newton though the
creator had put the beans (sun, planets, and satellites) in
the bag and given it a shake. But he thought the system
would lose its regularity, and after a few thousand years
the creator would have to give it another push. Laplace
showed that it would continue for at least 100,000 years
(a very long time in those days), and therefore informed
Napoleon that he had no need of the hypothesis of super-
natural interference. We still don't know if it is stable
for periods over 10^9 years, probably because we aren't sure
enough about relativistic corrections. I have got back to
beanbag genetics in a big way, largely because I have got a
colleague, Jayakar, who can correct my algebra. We find
that there are a whole lot of conditions other than superiority of heterozygotes which will conserve polymorphism on
a reasonable scale. 30 years ago I showed that mutation
would not, unless selective differentials were as small as
mutation rates, but that migration might do so. We can now
give the conditions as to migration rather more concretely.

The theory, which we are working out, of just what
happens when an initially "unfavoured" genotype gradually
increases its selective value is very tricky. It seems
that the population may change rather suddenly, even if the
relative fitness is only increasing slowly, and the selection
of "modifiers" may make this change still more sudden. Unfortunately even when selection is weak we need Bessel
functions, which occur in the theory of vibration of drums.
But we are beginning to see what may happen when climatic
conditions change slowly.

We have just worked out a fairly comprehensive theory

Contd...

: Page 2 :

of what happens under selection of constant intensity when
this is fairly strong (as it doubtless is when a new niche is
occupied and there is no immigration from the old one). For
this we need automorphic functions of a kind which were
fashionable in France about 1920. I may of course be hopelessly out-of date in my approach. I am sure bright boys
like Jim Crow think so. But it seems to me that mathematical
genetics are still about the stage of

$$s = \frac{1}{2} ft^2,$$

and that the mathematicians who come in from time to time are
interested in inessential, or shall we say, topics whose biological importance is not obvious. If I could have devoted
my life to the mathematical theory of evolution I might by
now be able to tell you a little more about populations. But
I have so much else to do, even as a mathematician, for example devising tests for hypotheses to be applied to samples
of a few hundred humans or birds. Salim Ali has data on
ringing and recapture which lead me to the estimate of about
80,000 migratory swallows in a roost in Bombay in winter.
Numbers in a roost of several _Motacilla_ species in Kerala
look like being a million or so. However these numbers tell
us little about population structure. We don't know where
these swallows breed (? Mongolia) or whether there is juvenile dispersal, even if adults have a fixed breeding place.

Jayakar and Spurway's observations on _Vanellus_ have
given me a hunch as to the origin of lactation which I will
send you when typed. There are about 7 steps, each useful
by itself. By the way you are a little unfair to "structure".
When, for example, the skin vessels take on the job of heat
regulation, they can only do it because they are "built-in"
structures in the central nervous system. We barely know
where they are, and nothing about their detail. But they are
structures all right.

I am a great believer in structure, but

he who mutation is to devised imperative of behaviour

Yours sincerely,

J.B.S. Haldane
(J.B.S. Haldane)

Prof. E. Mayr,
Museum of Comparative Zoology,
Harvard University,
Cambridge 38,
Mass.,
U.S.A.

Haldane

GENETICS AND BIOMETRY LABORATORY
GOVERNMENT OF ORISSA

No. 0402

BHUBANESWAR-3
ORISSA, INDIA

June 3, 1963

Dear Mayr,

If you read the enclosed article you will see that it is the answer to a challenge which you made in one of the reprints you sent me. It will, I hope, be published in the Rationalist Annual. If you think any of it worth reprinting, you may let me know who would be likely to reprint it. I don't think Evolution would take it in anything like its present form.

I have also completed a much more serious attack on you, entitled "A defense of beanbag genetics". This is intended for "Perspectives in biology and medicine". I may say that from defense I pass to counter-attack. However I am going to get it typed, and than look at it again after three months or so, to see if I find any sections of it unclear or unfair, or whether perhaps some new arguments have occurred to me. Here is an example of a counter-attack. On p.191 of "Animal Species and Evolution" you suggest that I did not believe in strong selection till 1957. In 1924 (Trans.Camb.Phil. Soc. recently reprinted by Comstock and Robinson of N. Carolina) I calculated that the mutant <u>carbonaria</u> of <u>Bisten betularia</u> conferred an advantage of about 50% . This was beanbag genetics, and nobody took it seriously for 30 years.

Jayakar and my wife are getting remarkable data both on wasps and birds. For the reason mentioned in the article, there are very few observations on animal behavior during the *be* | hot weather in the Indian plains, so they can fairly confident that most of their observations are new.

I may conceivably be in U.S.A. in October. Some of the N.A.S.A. boys have asked me, but I doubt if I shall get a visa. If I am there, I might be able to see you. But it is easier for you to get here. My house is about 1 km. from the place where you saw the larks in the early morning.

Dr. E. Mayr,   Yours sincerely,
Museum of Comparative Zoology,
Harvard University,   J. B. S. Haldane
Cambridge 38,
Mass.,   (J.B.S. Haldane)
U.S.A.

Enclr: Typescript of 'The origin of lactation'.

" -   " —all very mi—an

June 18, 1963

Dr. J. B. S. Haldane
Genetics and Biometry Laboratory
Government of Orissa
Bhubaneswar-3
Orissa, India

Dear Haldane,

I was delighted to get the copy of your manuscript on "The Origin of Lactation." I agree that this is a plausible explanation, although there are a few puzzling aspects which you do not mention. The Hairiness of mammals has always suggested to me (and to others) that the class may have originated in a temperate climate. It is of course possible that lactation preceded hairiness, but there is much to contradict such an assumption. The distribution of sweat glands in hairy animals is another matter to be considered. Horses, I believe, have a rather even distribution of sweat glands, dogs do not. The question of the salt content of sweat is another one which presumably enters the picture somewhere.

On the bottom of page 2, you might want to refer to various methods of spermatophore transfer in insects and urodeles, to the peculiar sperm transfer in some arachnids, and of course to the spawning in marine and other aquatic organisms, which is so similar to wind pollination.

On page 4, you might mention that some sand grouse seem to carry water in their crop, and of course you might refer to the crop milk of pigeons, which although fairly solid may well satisfy thirst for liquid. It is interesting that crop milk production is stimulated by the same hormone as milk production in mammals.

On page 6, I rather doubt that the Urey method will produce satisfactory results when applied to the reptile-mammal transition. Many, if not most reptiles have behavioral temperature regulation, and they are not "cold-blooded" at all. Indeed, by moving in and out of the sun, they manage to maintain a remarkably constant body temperature, which differs from species to species and from genus to genus.

The two species of Drosophila which Koopman bred in the same population cage coexist over large areas, in fact I do not know of a single place where you can catch only _D. persimilis_ and not also _D. pseudoobscura_. I have pointed out in my new book what keeps the two species from interbreeding.

Dr. J. B. S. Haldane - page 2   June 18, 1963

Lower down on the same page, I would place less credence on whatever Rhine publishes. No one knows what may or may not "occur" to Thine. I think the extremely careful work of Agar and Tiegs has certainly completely clarified MacDougal's experiment.

I shall receive your "attack" on me philosophically. In a big volume like the one I have written, it is quite impossible to avoid short-cuts and generalizations. For instance, I had your 1924 paper in an earlier draft, but took it out since you refer to it in your later papers, and I had to streamline my over-long bibliography. I wonder whether other readers would also come to the conclusion that I "suggest that[you] did not believe in strong selection until 1957." The whole point I tried to make was that around 1930, the emphasis was on the effect of slight differences in selection pressure and that this led many non-geneticists into making unrealistic assumptions.

I hope Helen and Jayakar are being helped with the literature on lapwings. There are some very interesting differences in courtship between different species and "genera," and a non-ornithologist might have trouble rounding up this literature. I should be glad to help if help is appreciated.

With best regards,

                                   Yours,

                                   Ernst Mayr

EM:ehw

Haldane

GENETICS AND BIOMETRY LABORATORY
GOVERNMENT OF ORISSA

No. 0473

BHUBANESWAR-3
ORISSA, INDIA

June 26, 1963

Dear Mayr,

Thank you very much for your letter. If I ever prepare a paper on the subject for a scientific journal, I will consider your points, some of which were, of course, known to me. Let me deal with them in detail.

of them
-u/

(1). I fully agree that hair in synapsids probably evolved in a cool climate. However presumably some spread to warmer climates, and I suggest that lactation developed among them. I am glad you think hairiness preceded lactation.

(2). Most carnivores have sweat glands mainly on their footpads. Dogs keep cool by producing a very watery saliva and a special type of shallow breathing to evaporate it. Cats seem to manage without any obvious mechanism. It is of interest that the gene <u>hairless</u> in mice suppressed lactation on its first appearance, though after some scores of generations in laboratories this character reappeared.

(3). I thought of dragonflies and so on. But the packet transferred always seems to be wet at least inside. Plants with **sticky** pollinia have gone back to some extent on the "dry" method.

(4). I shall be glad of the reference to sand grouse carrying water in their crops. I kept off them and pigeons because this is another way of arranging water supply. I know Riddle discovered prolactin in pigeons before it was known to act on mammals. Did you ever read O.W. Stapledon's " Starmaker" ? He describes a manlike race on another planet where the females fed the young pigeon-wise, and had large spout-like lips which were much admired if they conformed to the preferred pattern.

(5). I know lizards keep a steady temperature when they can. But ours, though most active in summer, eat a bit in winter, though they can't get above 20° . If they form bone in winter, there could be annual layers of different isotopic composition. A lot of our lizards, especially geckos, are crepuscular, and cannot regulate much by insolation.

(6). <u>Peccavi</u>. I thought the two species merely overlapped.

(7). I place no credence on Rhine. I want to point out, very politely, that he is a magician or a crass materialist according to the needs of the moment. Perhaps I was too polite. This often works in England, as when they say in

Contd...

: Page 2 :

Parliament "I respectfully suggest that my honourable and learned friend may have been guilty of a terminological inexacti-
tude", rather than "That little mister Jones is a bloody
liar".

(8). Your help with lapwing literature will be greatly
valued. Of course the nearest copy of any given journal may
be 1,000 miles away, or more. The main results of J. and S.
observations are probably statistical. They distinguished
A from B fairly easily, and saw enough copulations to be
clear that B was the male. So they will be able to compare
the times spent by the two spouses in incubation, and son.

Helen chronically makes affectionate remarks about you.
More important, she would like you (1) to tell Evans (wasps)
that she has sent him a letter C/o. the Scientific American,
and (2) to let her have Kenneth Cooper's present address.*
Their information on wasps is accumulating at a fantastic
rate. Already one can say that they have at least 3 patterns
of control of the sex of their offspring, all involving some
sort of hook-up of fertilization control and stimuli received
from receptors.

Yours sincerely,

J. B. S. Haldane
(J.B.S. Haldane)

Prof. Ernst Mayr,
Museum of Comparative Zoology,
Harvard University,
Cambridge 38,
Mass.,
U.S.A.   * He is a wasp-watcher and was, I think at Rochester

Spurway
Haldane

GENETICS AND BIOMETRY LABORATORY
GOVERNMENT OF ORISSA
BHUBANESWAR 3
ORISSA, INDIA

November 4, 1963

Dear Ernie,

This is
∧ just dashed off.   Thank you for interesting Evans, and for his book.

I am not in the least ironic about considering that only systematists have made lasting contributions to evolution theory. I have taught it to my students for fifteen years, and we have paused and considered what this infers. I do not think genetics has made <u>lasting</u> contributions <u>yet</u>. Any given "advance" is more likely to be stimulating than correct, or even to be the most efficient way of considering the matter. Also genetic erudition does not <u>in practice</u> seem to provide as fruitful a background for thinking about the matter, as erudition on the organic variation, not so much with in species, but at the species level.

Ernie you remind me of Bellaggio — when I do flatter you, you think I am laughing at you.

I will write again — this is not a reply.

Yours sincerely,

Helen
(H. Spurway)

Dr. Ernst Mayr,
Museum of Comparative Zoology,
Harvard University,
Cambridge 38,
Mass.,
U.S.A.

P.S. appologies for the italics *hv*

April 20, 1964

Professor J. B. S. Haldane
c/o Dr. Maynard Smith
British Museum of Natural History
Cromwell Road
London S. W. 7, England

Dear Haldane,

The news of your illness and operation reached me rather late, and then I seemed to be so immersed in travelling, lecturing, and whatnot, that I somehow did not manage earlier than today to send you my best wishes. In the meantime, I got hold of your "ode on cancer," which amused me with its Shavian sense of humor. I do hope that you are not too uncomfortable. Operations of carcinomas of the colon are sometimes amazingly successful, and I do hope that your case belongs in this category. I hope to pass through London this coming July, and will be very much looking forward to seeing you then, if you should still be in England.

With best wishes,

Yours sincerely,

Ernst Mayr

EM/ehw

HALDANE

GENETICS AND BIOMETRY LABORATORY
GOVERNMENT OF ORISSA

BHUBANESWAR-3
ORISSA, INDIA

No.64/0082.   May 2, 1964.

Dear Mayr,

Thanks for your letter of April 20th. I got back here
at the end of March, just in time for the hot weather. So
you won't see me in London in July. I am pretty well all
right, except that my colostomy has not learned to cope
with the diet here in a regular manner. This means that I
can't yet travel around much, a very great advantage if one
wants to work, as I do.

Jayakar and Spurway are making what I believe to be
important observations on birds. So far they have, I think,
identified 7% species in our garden, and perhaps as many in
the neighborhood. Observations include:—

1. Reports on nesting in <u>Vanellus malabaricus</u>. The
most novel observation is perhaps that the same pair of
birds did not wet their eggs in February, but did so in
April with water carried on the feathers. But they have a
lot on the division of labor between parents, and so on.
Publication in Rensch's Festschrift, Zool. Jahrbücher.

2. Dairy of presence of species. Apart from serious
migration, some species go from suburbs to forest in the
winter, and others in the summer. The distance is 5 to 10
miles. Is there <u>any good summary of this local migration</u>?
It is not vertical, as in the hills. Heights differ by under
100 meters.

3. Records of flight of <u>Bubulcus ibis</u> over the house
returning to roost, over 4 months or so. Relation of times
to sunset and weather conditions. Changes in mean flock
size.

Do you know of any records of this type other than
those given in Wynne-Edwards' recent book? These are
sketchy, compared with ours.

4. Observations on roosting of several species. In-
stability of some joint roosts.

If you can help Jayakar and Spurway with questions or
references they will be grateful.

You ask where I stay in London. With the Maynard Smiths,
with my sister, or in hospital. I do not know of any decent
hotels. The last one in which I stayed, owing to delay with

Contd...

: Page 2 :

an aeroplane, had thin walls between rooms, and my neigh-
bor kept his or her radio on till midnight or later. I
should think the best hotel in London is Brown's, near the
k/ Royal Society. It used to be full of decaying dukes. It
may cost a lot, but won't contain movie stars or any but
very superior con men. The management has probably not yet
heard of television. But they could cook. It is the kind
of place where Bertie Wooster's aunts stayed (if you read
P.G. Wodehouse).

Of course we do plenty besides bird watching. I do
beanbag genetics. Jayakar spent yesterday boring holes in
human | wood blocks, to verify that some wasp species lay female
eggs in big ones and male in small ones. Ajit Ray has and
abnormality (short 4th metatarsus causing short fourth toe)
which is very common here (about 1 per 1000). It is hereditary in a messy way. The gene concerned has a penetrance
under half. Very similar pedigrees are reported from Japan
and Yucatan, but not from Europe or U.S.A. I think this is
due to a difference in background. In Japan the main gene
affects the fourth finger in about 10% of cases. Here
only one in over 100 cases is so affected. In Europe it is
presumably recessive. This hypothesis seems to fit your
point of view fairly well. We have a lot of blood samples
which are being analysed for various characters in Europe.

I learn from the local press that I have been elected
to the National Academy. However I am awaiting a letter from
the Academy before I distribute cattle to 1000 brahmans.
The election can be explained on several grounds, e.g.

1. They though I was dying of cancer, and wished to
solace my last moments.

2. On my visit in October-November I produced something original in a lecture.

3. After your remarks, the beanbaggers felt something
must be done.

And so one might go on. Anyway I suppose I shall get the
P.N.A.S. free, which will be something. Ones first foreign
recognition, in my case by the French, is very gratifying.
But I now think it would be better to recognize younger men
than myself.

Dr. Ernst Mayr,   Yours sincerely,
Museum of Comparative Zoology,
Harvard University,   J.B.S. Haldane
Cambridge 38,   (J.B.S. Haldane)
Mass., U.S.A.

HALDANE

June 3, 1964

Dr. J. B. S. Haldane
Genetics and Biometry Laboratory
Bhubaneswar - 3
Orissa, India

Dear Haldane,

I was very pleased to get your welcome letter of May 2nd with all of its good news. No doubt the hot season is going to be a trying time for you, until your system has become entirely adjusted functionally.

We were all saddened to learn of Nehru's death. Although he had his weak points, as any major figure, he nevertheless had an enormously stabilizing effect on India and is perhaps more responsible than anyone else for the determination of India to move forward in spite of all difficulties.

The editor of Perspectives in Biology and Medicine has asked me to respond to your statement on beanbag genetics, but I am not sure that I will do so. Obviously most of what you say will be fully endorsed by me also. It is all a matter of emphasis. A certain amount of beanbag genetics is the necessary basis for all else, but on the whole, beanbag genetics is singularly unsuitable to explain any but the most elementary evolutionary phenomena. Penetrance, as you rightly remark in your letter, is a case in point. The recent models developed by Jacob and Monod, and others, about regulating genes are further substantiating arguments. The enormous increase in the amount of DNA among the eucaryotes, most of it apparently not used for structural but for regulating genes, is still further evidence. It is no use to argue about trivia, but I still believe that beanbag genetics in many cases had a detrimental rather than beneficial influence on evolutionary thinking. You, yourself, without using such terminology, have called attention to this in many of your papers.

I realize that your election to the National Academy is not an honor to you but a removal of a disgrace from us. We have an excessively democratic system of voting in ordinary members, but Foreign Members are proposed by the Council, and this means that the influence of the biologists is rather limited. I do not know whether this will please you or not, but I can assure you that you have been for years the leading candidate of the biologists. It has been a great source of satisfaction for all of us that your election was finally ratified. Hence even if the election should not mean a thing to you, I can assure you that it means a lot to us.

Dr. J. B. S. Haldane - page 2   June 3, 1964

     I will write to Helen myself with respect to her bird observations. Much as we know about birds, we really know very little when it comes down to genuine detail. There is an enormous amount of literature about the relation of roosting flights to time of sunset. Otherwise roosting phenomena have a strongly local character. Roosts may exist for many years, or shift from month to month, etc.

    With best wishes,

               Yours sincerely,

               Ernst Mayr

EM/ehw

HALDANE

GENETICS AND BIOMETRY LABORATORY
GOVERNMENT OF ORISSA

TELEPHONE: 78
TELEGRAMS: ELAH

BHUBANESWAR-,
ORISSA, INDIA

September 29, 1964

Dear Mayr,

I must first thank you for your present. I regard pages 285-287 as perhaps the most important three in the book. For the first time Darwin gave a nearly correct geological time scale. As you know he was persuaded to withdraw them by kind friends. The same with the bear and the whale story on pp.

Of course I agree with most of your letter to Helen of September 23rd. Biswas and Salim Ali can and do help us. But as Jayakar has now identified 79 bird species in or over our garden he does not need a lot of help from them on this matter, whereas Helen and he can both help them with statistics and the presentation of numerical data. Again I have not yet seen what I regard as an adequate statistical treatment of roosting times in relation to sunset. If you know of one, give her the reference. Helen and/or Jayakar can tell you about immediate pre-roosting behaviour. I know Ford's butterfly stories fairly well. I have read his recent book.

As for my condition, it is not as good as I hoped. Owing to gross and systematic lying by English medicos I now learn that, instead of convalescing, as they said, the cancer is still there and spreading. Within a week a large silver lining appeared. A Calcutta biochemist and physician has developed a treatment which certainly checks growth and improves symptoms in some human cases of inoperable cancer (almost all Indian cases are inoperable before, if ever, they even consult a surgeon). I have put myself in his hands. If he fails with me that is a fact for his statistics. If he succeeds, I can give him a real boost. I don't mention his name, because several highly placed Indians have already done their best, often successfully, to handicap his work, which is based on a "new perspective", neither ~~based~~ on Western nor Indian practice.

So if you learn that I am dying of cancer, tell them to wait a year or two for verification. Don't, in principle, write to me, as I am still, and perhaps permanently, rather weak.

Dr. Ernst Mayr,
Museum of Comparative Zoology,
Harvard University,
Cambridge 38,
Mass',
U.S.A.

Yours sincerely,

J.B.S. Haldane
(J.B.S. Haldane)

October 21, 1964

:. Helen Spurway
metics and Biometry Laboratory
vernment of Orissa
ubaneswar-3
rissa, India

ar Helen:

   Needless to say I was saddened by Haldane's letter of
ptember 29. One always hopes against odds that a cancer operation
s a radical success, and even if it is not, there are enough cases
f inexplicable remissions to maintain ones hopes. I see nothing
rong or ridiculous in trying the treatment suggested by the Calcutta
iochemist.

   I still remember the Pavia meeting, not because Haldane
lmost choked with a chicken bone in his throat, or because of his
ncredible virtuosity with verses from Dante's Comedia Divina, but
ecause Haldane was the only one of the older geneticists present
ho fully appreciated what the younger ones were talking about.
. A. Fisher disappointed me at the time by simply resisting the new
deas. Haldane accepted them at once (so far as they were sound)
nd asked meaningful questions as to where to go from here. This
_exibility of his mind no doubt is the reason for his many original
deas. His 1957 paper on the cost of natural selection is another
ocumentation of his continuing originality. By the way, I hope you
an discuss with him the paper by Alice Brues in the September issue
f Evolution, page 379. She makes some good points there supplementing
n part the questions I asked on pages 260 and 261 of my book. I
hink this will be the starting point for some interesting future
levelopments.

   I wish I could help you better with the literature on
oosting and sunset. There is a lot of such literature, but I do
ot have any assistant right now who could dig it out. I hope to
'ind someone in the near future. ⟨ ⟩-Jayakar researches have made
i little progress, but I may be able to find out still more.

   Please give Haldane my best wishes. He asked me not
to write to him directly and so I am writing to you.

         Yours ever,

         Ernst Mayr

EM:sm

MUSEUM OF COMPARATIVE ZOOLOGY
The Agassiz Museum

HARVARD UNIVERSITY
26 OXFORD STREET
CAMBRIDGE, MASSACHUSETTS 02138

31 July 02

207 Badger Terrace
Bedford MA 01730
But address

Dear Krishna,

Thank you for your letter of July 22.

I am delighted you are doing a biography of Haldane. I always tell everybody that Haldane was the most brilliant person I ever met in my life; considering his enormous gifts it has always been a puzzle to me that he is not made more great discoveries that would be recorded in a history of biology. I think he had too many interests and never concentrated on any specific problem in evol. biology.

Of course, I feel honored that you plan to publish some of my letters and I herewith authorize you to reprint them or parts of them. Since Haldane's thinking was in some respects so similar to mine, it was a great loss for me that he died so prematurely. I also authorize you to select the portions of my letters which you want to publish.

Don't forget to mention what an extraordinary interest in natural history Haldane had, as I discovered on my excursion with him in Berhampur or so.

With best regards
Yours

Ernst Mayr

Index